YOU ARE HERE

A User's Guide to the Universe

Richard Farr

Also by Richard Farr:

Emperors of the Ice:
A True Story of Disaster and Survival
in the Antarctic, 1910-1913

The Truth about Constance Weaver

The Fire Seekers

www.richardfarr.net

In memory of my grandfather, H.C. 'Jimby' Reynard—Wagner fan, soldier of the Great War, scientist. On my tenth Christmas Day he gave me a slide rule. All that afternoon, we sat on the carpeted bathroom floor, with our backs against the radiator because it was the quietest warm place in a snow-bound house stuffed with relatives, while he patiently taught me its secrets.

And with many thanks to John Harvey, who labored heroically, like Hercules in the stables of Augeas, to cleanse the manuscript of problems digital, typographical, grammatical, and factual.

*The beginning of every science is the description
and naming of phenomena.*

— E.O. Wilson

*Where the telescope ends, the microscope
begins. Which has the grander view?*

— Victor Hugo

*If at first an idea is not absurd,
then there is no hope for it.*

— attributed falsely, but with
delicious plausibility, to Einstein

TABLE OF CONTENTS

Introduction: Welcome Home

A noise wakes you up. Leaves rustling outside your window? You keep your eyes closed for a few seconds, pull the blanket tighter around your shoulders, and savor the sensation that you're safe, warm, at home in your bed.

But then your brain registers that something is wrong. The rustling you heard is not the sound of leaves, but of voices. And this is not the edge of a blanket in your hand, but the collar of a coat. Beneath you are hard wooden slats.

You sit up and open your eyes. Now you discover, to your horror, that you are not at home at all, not in bed, but sitting on a bench, at the edge of a park, in an alien city.

Alien.

Everything single thing your senses reveal to you is unknown, from the floating buildings and the five-eyed people to the three reddish moons in a grass-green sky.

You have a choice now. You can shut your eyes again and hope that this strange, frightening vision will melt away like a dream. Or you can accept that it's real, and start exploring.

Might as well get up. Whether you like it or not, somewhere utterly alien and unexpected is where you are, right now. To find out just how alien and unexpected, you only have to open your eyes and let science guide you around the world you thought you knew. Science fiction may impress or amuse us with its inventions, but science reveals an actual world that is, as Arthur C. Clarke once observed, far stranger.

What science has to say about our alien world belongs to all of us, and it can be imagined and grasped and understood. One motive I had for writing this book is the sheer irritation I feel when people say lazily that this or that thing is "just too big (or small, or complicated, or weird) to imagine"—as if that's a good excuse for not bothering to try. Maybe some things are impossible to imagine, but let's not give up so easily. Let's see if we can find ways to imagine them. Because science tells us things about where we live that are, in the most literal sense of that overused word, *wonderful.*

This book is a tourist guide to your universe. A hit-the-highlights introduction to *everything,* it aims to give you a sense of what our world's furniture is, and how it's bolted together. The journey goes in steps, like rungs on a ladder: upwards, to the very large scales of geography and astronomy and, at the same time, downwards or inwards to that other great space, the limitless province of the microscope. At each step you are ten, a hundred, ten trillion times removed from the human scale—*in both directions at once.* So it will be apparent at a glance that bacteria are to you as you are to a continent. Or that you are to Mount Everest as Mount Everest is to Saturn.

In science, being a nut for accuracy is usually a good thing. The cesium fountain / ytterbium atomic clocks at the US National Institute of Standards and Technology and the UK's National Physical Laboratory are among the best timekeepers ever made, accurate to better than a second every 100 million years. This kind of accuracy isn't just for nerdy bragging rights—it has made possible the most precise length measurements ever. By clocking laser pulses as they bounce back from three reflectors left on the lunar surface by the Apollo astronauts, we can tell the distance to the Moon within a centimeter. Because of that, we're in a position to test ideas about gravity, seismology, and other subjects. That umpteenth decimal place can make a difference. In this book, though, I'm not interested in extreme precision. I'm interested in helping you *get the idea*. Suppose I tell you *The Moon's orbit has a radius of 3.5641 x 10⁸ meters at perigee*. It's true! But even if you can read "power numbers" confidently, and know that "perigee" is the point in the Moon's orbit that brings it closest to the Earth, this is useless, because it gives you no image: you have information without understanding. It's much more useful to know that the Moon is:

- a quarter million miles away, or (better)

- thirty times farther away than the Earth is wide, or (better still)

- **like a sheep a hundred paces from an elephant.**

So I've rounded many of the numbers in this book, used the word 'like' a lot, and focused on helping you get a clear *picture* of the relationship between different scales.

I describe some remarkable discoveries here. But none is more amazing than this: just a generation ago, in a momentous epiphany that still has barely begun to trickle down into the popular consciousness, scientists grasped for the first time that these scales—stretching from the cosmic to the sub-atomic—are not so much a line as a circle. **It turns out that the largest and smallest scales link up, and make one another meaningful, in a manner so beautiful and so profound that no storyteller could have invented it.**

But we're jumping ahead. Before we get to the end(s) of our journey, the ends that seem so far apart until they magically join, we must traverse many orders of magnitude.

[WAIT. ORDERS OF WHAT?

An order of magnitude is just a multiple of ten. The largest whales and dinosaurs are 30 meters long: that makes them an order of magnitude larger than horses and zebras (3 meters long), two orders or magnitude larger than rabbits (30 cm long), and three orders of magnitude larger than most insects (3 cm or so). So 'orders of magnitude' are just a useful guide to size. It takes three of these scale changes to take us from the whale to the beetle. It takes at least *sixty two*—arguably, far more—to take us from the largest to the smallest things we know. As we'll see, some rungs of this universal ladder are crowded with stuff, while others are spookily empty.

But the main (and easy) thing to remember is that 10^3 means 10 x 10 x 10, or 1,000—so a power of 3 or -3 is "thousand" or "thousandth," a power of 6 or -6 is "million" or "millionth," and a power of 9 or -9 is "billion" or "billionth."

Powers are powerful: 10^6 seconds is less than two weeks, 10^9 seconds is more than 30 years, and 10^{12} seconds takes us back to the extinction of the Neanderthals. And compare these two numbers: 1.27×10^7 meters ("tens of millions") is the diameter of the Earth. 1.27×10^{-7} meters ("tenths of millionths") is the diameter of a virus.]

We begin in the middle of these many magnitudes, because something in the middle is familiar to you: you. Or let's put it more bluntly. What's most familiar to you, because you're it, is a very bizarre bipedal life-form made up (like its environment) mostly of an oxide of hydrogen. This creature can't survive even for a few minutes without inhaling a gas, oxygen, which scarcely existed on the early Earth and was a deadly poison to the Earth's earliest inhabitants.

Luckily for us, ancient volcanoes belched oxygen into

the atmosphere—and, later on, blue-green algae (cyanobacteria) belched even more of it. Not everyone whooped for joy: oxygen is so corrosive that it eats metal, and it drove Earth's earliest inhabitants, the bacteria-like Archaea, deep underground. To the Archaea, our world must look like the aftermath of a nuclear holocaust, with the entire surface of the planet irretrievably poisoned. "It was," they are probably saying to their children right now, "not meant to end this way." But their near-end was our beginning.

Homo sapiens is a very, very recent arrival on this scene. Writer John McPhee coined the term Deep Time and thought up the idea of comparing your arm to the history of life on Earth. Stick one arm out in front of you, fingers extended. Now swipe a nail file across the fingernail of your middle finger. If the length of your arm is the history of the planet, the few particles of fingernail you just erased are the entire history of our species.

We are confined in space just as severely as in time, inhabiting only the driest, warmest few per cent of the surface of a small-to-medium-sized rocky planet ("Earth") orbiting an unremarkable yellow star ("the Sun") in the outlying suburbs of a medium-sized spiral galaxy ("the Milky Way") in the outlying suburbs of the inconspicuous Virgo Cluster of galaxies.

Among other dramatic changes in perspective, we have had to accept that:

- Even the Earth's sun is not the center of the universe;

- All the couple of thousand stars we see from Earth on a clear night—away from the light-polluted skies of our cities—are just a tiny fraction of all the stars in the Milky Way; not one in a hundred, not even one in a million;

much, much less than the proverbial drop in a bucket;

- Even the few hundred stars nearest the Sun sport dozens of other solar systems;

- As the philosopher Immanuel Kant guessed back in the eighteenth century, the Milky Way itself is just an "island universe," one little speck in what turns out to be an archipelago of a hundred billion islands spread across the universe like silver dust on a black silk sheet.

- We evolved from other life-forms—and not even, as pious Victorians feared, from monkeys or apes, but from something a lot more like bacteria.[1]

---

[1] **NOT EVEN FROM MONKEYS OR APES:** Your great-great-grandmother was not a gorilla. The apes and monkeys are our *cousins*: not the trunk from which we branched, but branches off the same trunk. 7 or 8 million years ago, creatures existed that were probably a bit monkey-like and a bit human-like; they don't exist anymore, but some of their descendants evolved into gorillas (one or arguably two species surviving), others into chimps (two species), and others into humans (one species surviving; several others, including the Neanderthals, the Denisovans, and probably more, extinct). So no gorilla is your ancestor, but every gorilla is roughly your quarter-of-a-millionth cousin. For the full gloriously messy story of how we and our ape-cousins emerged from an earlier world of prokaryotes, sponges, and flatworms, see Richard Dawkins, *The Ancestor's Tale*.

There's more yet—we have barely mentioned our even more recent and equally incredible discovery of the seething, almost incomprehensibly huge spaces that open up to us under the microscope and the particle accelerator.

We now set out to walk over all this ground, pointing out the main sights. But "set out" gives the wrong impression. When we open our eyes, dust off our trousers and go exploring in this alien landscape, what we are really doing is coming home. T.S. Eliot was no doubt thinking of human life in general, but might easily have been thinking of science, when he wrote:

We shall not cease from exploration
And the end of all our exploring
Will be to arrive where we started
And know the place for the first time.

Know the place: now there's a goal everyone should reach for. A universe is a terrible thing to waste.

Scale Zero: Your Neighborhood

$10^0 = 1$, so 10^0 meters is one meter: *your* scale.

Almost all humans start life about half a meter long, and stop growing between 1.5 and 2 meters; even as an adult you're a meter-sized creature sitting down.

The shortest fully-grown person who ever lived was probably Gul Mohammad of Delhi, India. On July 19, 1990 he was just over half a meter tall.

There's some variation in typical height among different human ethnic groups, but children vary less than adults. A typical healthy girl of (say) northern European ancestry reaches 1 meter at about the age of four.

Africans have more genetic diversity than all other humans put together, so it's not surprising that it provides both the shortest ethnic group in the world (the Mbuti or Bambuti people of eastern Congo, with an average adult height below 1.4 meters), and the tallest ethnic group (the Dinka of southern Sudan, who are often closer to 2 meters; basketball star Manute Bol, a Dinka, was 2.3 meters or 7 feet 7 inches).

Humans share the 1 meter scale with many other creatures in the air and on land. An adult bald eagle (*Haliaeetus leucocephalus*) is 1 meter from head to tail with a 2 meter wingspan; surprisingly, the flying fox *Pteropus vampyrus* is about the same size. Africa's marabou stork (*Leptoptilos crumeniferus*) has a wingspan up to 3 meters; the wandering albatross (*Diomedea exulens*) has the largest wingspan of any living bird—over 3 meters.

Amazingly, the world's longest stick insect is in our range: in Borneo and Malaysia you might run into a couple of species from the genus *Phobaeticus* that are over half a meter long including their outstretched legs.

It may come as less of a surprise that many fish are as big as us—but some of them are really, really big. The Atlantic tarpon (*Megalops atlanticus*), Napoleon or Humphead wrasse (*Cheilinus undulatus*), Amazonian giant arapaima (*Arapaima gigas*), and electric eel (*Hypopygus lepturus*) can all reach 2 meters or more. The Atlantic halibut (*Hippoglossus hippoglossus*) can reach 3 meters. This is about the same length as the Komodo dragon (*Varanus komodoensis*) of Indonesia's Lesser Sunda archipelago.

Oh, and welcome to Australia ...

The Lion's Mane jellyfish (*Cyanea capillata*) is the world's largest at 2 meters across—and has tentacles that dangle down 40 meters. The box jellyfish *Chironex fleckeri* isn't quite that big—its bell might be the size of your head, with tentacles a couple of meters long—but it's worth mentioning as one of the most lethal animals on the planet: its millions of stingers or nematocysts can deliver a toxin direct to your bloodstream that causes unbearable pain followed by a heart attack. Get out of the water and run away! Run all the way to the Australian outback, where you might be bitten by a 2-3 meter inland taipan (*Oxyuranus microlepidotus*), the most poisonous snake in the world.

One of the most remarkable organisms at our 1 meter scale is the world's largest flowering plant, Titan arum (*Amorphophallus titanum*), native to Indonesia. The "corpse flower" blooms only for a day or two every few years—which is a good thing, since the "scent" (as I can tell you from first-hand experience) is a combination of rotted meat and vomit with undertones of gasoline. The bloom is several feet wide, and surrounds an erect central section, the spadix, that's about 8 feet tall.

Among the primates, the gorilla and orangutan are about our size. Among mammals, we are on at least roughly the same footing (so to speak) as red kangaroos (*Macropus rufus*), medium-sized bears such as the American black bear (*Ursus americanus*), seals, dolphins and dugongs, the guanaco (*Lama guanicoe*), the mountain tapir (*Tapirus pinchaque*), the okapi *(Okapia johnstoni)*, and the pygmy hippopotamus (*Choeropsis leberiensis*)—to name a few.

Excluding its tail, a domestic cat (*Felis catus*) is about half a meter long; its largest cousin, the Siberian or Amur tiger (*Panthera tigris altaica*) can be over 3 meters long.

There are thousands of other plant and animal species at our scale, but they're relatively familiar. So—with a nod to the fact that we are also surrounded by an invisible ocean of microwave and FM radio waves many of which are about our size—let's move on up, and down.

Scale One: Of Mice and Mammoths

$10^{-1} / 10^{1}$

TEN CENTIMETERS / TEN METERS

TEN TIMES SMALLER AND LARGER

A pygmy mouse lemur is to you as you are to a whale shark.

In our immediate evolutionary family, the primates, the smallest member is the pygmy mouse lemur *(Microcebus myoxinus)* from the forests of Madagascar. It's just over 6 centimeters (2.5 inches) excluding its tail, and weighs 30 grams—the same as a piece of toast.

The Cuban bee hummingbird, *Mellisuga helenae*, the smallest of all birds, is not much smaller than the pygmy mouse lemur, but at 2-4 grams it weighs much less.

Birds are usually far lighter than they look, partly because evolution had the clever idea of hollow bones. (This isn't the whole story: birds need some leg and chest bones to be very strong, and these can be relatively heavy. At least as important is having a small, light skull without the baggage of mammalian jaws and teeth.) Hummingbird moths (family *Sphingidae*) are appropriately named, as they are sometimes mistaken for hummingbirds; they are about the same size, but the moth is heavier than the bird.

Also at this scale are the world's smallest monkeys: pygmy marmosets of the genus *Cebuella* are typically 15 centimeters (6 inches) long with a 23 centimeters tail. A much stranger-looking animal—partly because it does no looking—is the 10 centimeter Texas blind salamander, *Eurycea rathbuni*, one of the hundreds of cave-living species that obviously once had eyes but have lost them.

Another primate, Brazil's Golden Lion tamarin *(Leontopithecus rosalia)* is a relative giant at 25 centimeters (10 inches) and half a kilogram (a bit over a pound). That's the size of a small Norway or brown rat (*Rattus norvegicus*)— and also, even more alarmingly, of the world's largest spiders, the heavy South American bird-eating species *Theraphosa leblondi* and the smaller-bodied but even longer-legged Laotian giant huntsman (*Heteropoda maxima*), which can reach 30 cm (12 inches) across.

All these animals could easily fit inside the world's largest seed, the Coco de Mer (*Lodoicea maldivica*), which can weigh as much as a three-year-old child and be 30 cm long. Just a few of the non-primate mammals that could also curl up comfortably inside a Coco de Mer seed include:

- most of the more than 50 species of weasels and their relatives (Family *Mustelidae*, including the polecats, stoats and ermines)
- most of the more than 30 species of mongoose (Family *Herpestidae*)
- all of the civets and linsangs (30 species in the Family *Viverridae*)
- either species of the amazing colugos (Family *Cynocephalidae*, perhaps the best gliders among all the non-flying mammals)
- most of the more than 70 different species of opossum (Family *Didelphidae*).

As with the spider, it's only the freaks of the insect world that make it to this scale: the world's heaviest beetle, for example, *Goliathus goliatus*, which can be 11 centimeters—the right size to fill your hand.

Strangest natural phenomenon at the 10 centimeter scale? Maybe it's nature's largest eye, which belongs to the Collossal squid (genus *Mesonychoteuthis*). 'Giant' and 'Collosal' squid can measure more than 12 meters long, and a 40-foot squid would have an eye about 30 centimeters in diameter. As far as we know, these are the largest eyes that have ever evolved. Their owners are a bit of a mystery; probably the largest living invertebrates, they live in deep water, only occasionally wash up heavily decomposed, and have only very briefly been filmed—never captured and examined—alive.

The eel-like oarfish (*Regalecus glesne*) can reach 8-10 meters. The largest living reptile is the estuarine or saltwater crocodile *(Crocodylus porosus);* males average 4.5 meters, but at least one specimen was over 8.5 meters. Some extinct crocodilians, such as the 110 million year old *Sarcosuchus imperator*, which cruised long-gone waters in what is now the Sahara desert, may have reached 12 meters—the length of a bus. 12 meters is also the length of the world's largest surviving fish, the rare whale shark (*Rhincodon typus*).

Everyone knows about tape-worms, but we tend to think of them as one species—like koalas, only less cuddly—when in fact there are thousands of species. (One lives exclusively in the guts of bearded seals.) The broad fish tapeworm (*Diphyllobothrium latum*) can grow in your very own gut for 20 years, where it may reach a length of 10 meters. Every felt like someone inside you is constantly hungry?

The world's largest species of cactus is the Sonoran desert's saguaro (*Carnegiea gigantea*); typically 7 meters (23 feet) at maturity, it can grow to over 15 meters (50 feet).

The fin whale, *Balaenoptera physalus*, and the extremely rare bowhead whale (*Balaena mysticetus*) can reach 20 meters. But the largest animal is of course the blue whale—also known as the sulphur-bottom whale because it tends to collect yellow diatoms on its underside. The female of the larger southern sub-species *Balaenoptera musculus intermedia* can grow to 25-30 meters and 85-100 tons.[2]

(People tend to exaggerate the size of the blue whale. Specimens over 33 meters have been reported in the past, but only one individual was ever reliably measured at that size. After centuries of exploitation it's almost certain that no individuals so large survive. The species is in extreme danger of extinction: from an historical population of perhaps half a million, fewer than 5,000 remain in the world.)

The largest land animals ever were the sauropods, plant-eating dinosaurs that lived between 165 million and 65 million years ago. *Argentinosaurus* may have been almost as big as the blue whale, at 30 meters long and 70 tons; a similar but even larger species, discovered in 2014, may have been 77 tons and as tall as a seven-story building.

---

[2] Strictly speaking, "ton" can refer to the American "short ton" (2,000 pounds), the tonne or "metric ton," (1,000 kilograms or 2,204 pounds), or the British "long ton," (2,240 pounds). I have used the short ton for most of my sources, but the long ton is only 12% bigger, so these round number would be about the same in US, metric, or British units.

Scale Two: Tiny Frogs and Giant Fungi

10^{-2} / 10^{2}

A CENTIMETER / A HUNDRED METERS

A HUNDRED TIMES SMALLER AND LARGER

The world's smallest fish is to you as you are to a giant sequoia.

The only mammals to have mastered flying are bats—but there are a thousand species of bats. The smallest bat (and the smallest mammal) in the world is Kitti's hog-nosed bat *(Craseonycteris thonglongyai)*, from western Thailand. It's also known as the bumblebee bat because it is only 30 mm (just over an inch) from head to butt. Fully grown, it weighs less than 2 grams—a tenth as much as a mouse.

All those bats make up roughly a fifth of all mammal species, and mammals make up about a fifth of all tetrapods (four-legged vertebrates). The smallest of all tetrapods are various species of Cuban and Papuan frog that measure less than a centimeter (less than half an inch) from snout to vent. This is about the size of the world's smallest fishes, such as the "stout infantfish," *Schindleria brevipinguis* (a scaleless and transparent inhabitant of Australia's Great Barrier Reef), which weigh about a thousandth of a gram. (You'd need to fry half a million to make a single meal.)

There are countless billions of insects in the world; they are common at several scales, but the 1/100 meter (1 cm) scale is where we are most familiar with them. There are so many insects that their total weight far exceeds that of all humans. Even the world's ants alone weight more than we do. How many species there are is really anyone's guess: more than 12,000 are known.

And then there are the astonishingly diverse *Coleptera*, or beetles. (The name means "sheath-winged".) There are at least 350,000 species—a third of all insects. Biologist J.B.S. Haldane was asked what our knowledge of nature could tell us about the nature of God. "He has," he replied, "an inordinate fondness for beetles."

By the way, most people think—and some science books repeat—that all six-legged animals are insects, but the millimeter scale is home to three classes of "non-insect hexapods." These creatures, the springtails, proturans, and diplurans, have six legs, but are classed as non-insects because their bodies are so different from classic insect bodies. These 'mesofauna' (literally: middle-sized creatures, that is, not macro and not micro either) are divided into over 30 families totaling nearly 8,000 species. In some soils they exist at densities of hundreds of thousands per square meter.

The largest living things are plants. The largest of all may be the giant sequoia of the Western U.S., *Sequoiadendron giganteum*. "General Sherman", a tree in California's Sequoia National Park, is 85 meters (nearly 300 feet) high and weighs about 2,000 tons.

Taller than *Sequoiadendron giganteum* but less massive is *S. sempervirens*, of which the tallest living examples (all in northern California's Humboldt County) are over 110 meters. Examples of this species—and of both the Australian eucalyptus (*Eucalyptus regnans*) and Douglas fir (*Pseudotsuga menziesii*)—have almost certainly exceeded 120 meters in the recent past. An Australian eucalyptus observed near Watts River, Victoria, in 1872 was believed to have been *over 150 meters* (nearly 500 feet).

Think we must be finished with living things? Well, it depends what you mean. Some quiet, unassuming monsters require us to wait for the next scale up ...

Scale Three: "Animalcules" and Canyons

$$10^{-3} / 10^3$$

A MILLIMETER / A KILOMETER

A THOUSAND TIMES SMALLER AND LARGER

"Animals" too small to notice are everywhere—from the desert to the Arctic, from your eyebrows to your underwear. Just take one example from the many that most people have never heard of: the phylum *Tardigrada*, or "water bears." Typically half a millimeter long—the size of a sugar grain— they live in hot springs, the Arctic, and your glass of water. They don't do a whole lot even fully awake—*tardigrade* means "slow walker"—but their cleverest trick is to slow down even more, entering a state known as cryptobiosis, in which their water content falls from 85% to 3%. In this form they can survive for years in temperatures up to 150 degrees Centigrade and down to -272 degrees Centigrade. There are several hundred species (some authorities say as many as 750) divided into around 100 genera. Humans may not sur- vive your favorite Hollywood disaster—earthquake, nuclear war, alien invasion, zombie apocalypse—but you can bet the water bears will carry on proliferating as if nothing had happened.

There is an immense zoo of life-forms at or around this scale, including:

- Gnathostomulids and platyhelminths (two classes of worm that live between grains of sand on the sea bed, where they eat bacteria

or the body parts of other small animals)
- Thrips (at least 5,000 species)
- Lice (at least 3,500 species)
- Fleas (at least 1,800 species)
- Ticks and mites (at least 32,000 species)
- The smaller wasps; those of the genus *Trichogramma*, which have the nasty habit of laying their eggs *inside* the eggs of other insects, are typically a bit more than half a millimeter.

Humans are all too cozily familiar with some of these: the human flea (*Pulex irritans*) is typically about 3 mm long, as is the head louse (*Pediculus capitis*). The very small tick *Ixodes ricinus*, about the size of a sesame seed, likes to bite you too. When it does so, you get an injection of the bacterium *Borrelia burgdorferei*. This little creature just loves your bloodstream, where it multiplies like crazy; you then get Lyme disease.

Large mites are 1-2 mm—the size of a single sugar grain. If you shrank to the size of a mite, a large temperate-latitude house spider would seem to stand on legs taller than most trees, with a body the size of a whale.

Dust mites live in every house and every vacuum cleaner, at typical concentrations of a thousand per gram of dust—think of each gram as a crowded medium-sized concert hall, and a full vacuum cleaner bag as a large, packed sports stadium. Go mites!

One of the commonest creatures on the planet is the roundworm or nematode. About 20,000 species have been described, though some scientists think there may be half a million; a bucket of garden soil can contain a million individuals. They include the parasitic hookworm, a major cause of disease in humans.

The tiny crustaceans known as copepods, a key element in ocean food chains, are typically about 1 mm.

Biologists group animals by kingdom, phylum, class, order, family, genus, and species. There are millions of species, and new ones show up all the time, but a creature strange enough to put into a whole new genus or family is much more unusual. In 1995, scientists announced that they had found a species too bizarre to fit into any of the 35 or so known phyla. That means it was so weird, so unlike anything else, that about all they could say was, "Well, it's not a plant or a fungus; it's ... it's ... a sorta kinda animal, I suppose." They were looking at *Symbion pandora*, a 0.3 mm parasite that lives exclusively on the lips of *Nephrops norvegicus*, the Norway lobster. Its very own phylum, *Cycliophora*, means "ringed creature" and comes from its funnel-like mouth parts. Its utterly bizarre lifecycle and reproductive system have been described as trisexual; it's more accurate to say that it has several different life 'stages' including an asexual feeding stage that can itself produce asexual stages, males, or females. The asexual stage can also 'self-renew'—essentially, it wears out and grows a brand new version of itself from within.

Typical bacteria are still two whole orders of magnitude away—but, as elsewhere in nature, there are freaks. The largest known bacterium, *Thiomargarita namibiensis* or "Pearl of Namibia," is visible to the naked eye at over half a millimeter. This is several hundred to a thousand times the size of typical bacteria—it's like average humans standing a meter and a half tall, but a few of us being twenty times the height of the Statue of Liberty.

The part of the iceberg that ripped open the Titanic was under water, like 70-90% of any iceberg. One monster berg, first spotted off the coast of Greenland in 1957, projected 170 meters (550 feet) out of the water, the height of a 50-story building—and thus was presumably at least 700 meters (2,300 feet) 'tall' overall.

If you shrink the tallest recorded iceberg until it's the size of a blue whale, the whale is the size of a toy poodle, and you are the size of a hummingbird.

And now for the biggest living things of all—at least on this planet. Specimens of the soil fungi *Armillaria gallica* and *Armillaria ostoyae* in Michigan, Washington and Oregon cover dozens of acres of forest and are thousands of years old. The one in Eastern Oregon covers about ten square kilometers, and is at least 2,400 years old.

In the Wasatch Mountains of Utah there is a 106-acre stand of genetically identical quaking aspen (*Populus tremuloides*). They don't cover as much ground as the fungi, but their 47,000 genetically identical trunks weigh over 5,000 tons. The aspen clone may be more than 10,000 years old.

Venezuela's Angel Falls (named after American aviator Jimmy Angel, but known to the Pemones Indians as Churún Merú) is the tallest in the world. It's 20 times as high as Niagara, and descends almost exactly 1,000 meters from the Auyantepuy (Auyan tableland).

You are to Angel Falls as a mite is to you.

Other interesting features of the natural world at the kilometer scale tend to be a matter of local pride and much dispute. Hawaiians think the world's highest sea cliffs are on the north shore of Molokai, and typically give the height as

600 meters. The Taiwanese claim their highest sea cliffs top this at 800 meters. But then some sources measure the highest point of the Molokai cliffs as about 1,000 meters.

So with the deepest gorges and canyons: and here, even more, it depends what you measure. At least five gorges on three continents can lay a claim. Obvious contenders include the Kali Gandaki gorge in Nepal (between the mountains Annapurna and Daulaghiri) and Hell's Canyon, on Idaho's Snake River in the United States. But they are probably beaten by Peru's Colca Canyon, outside Arequipa, which can claim a consistent depth of over 3,500 meters (11,500 feet). And figures over 5,000 meters (16,500 feet) have been claimed for the remote Yarlung Zangbo or Tsangpo Gorge in south-eastern Tibet.

The Greenland ice cap is about 3,000 meters (3 kilometers or two miles) thick at the middle. Most of the land beneath it is below sea level; if the ice melts, Greenland will end up looking like a big semicircle of islands, with a giant lagoon in the middle facing Canada.

Scale Four: The Smallest Insects and the Largest Mountains

$$10^{-4} / 10^{4}$$

1/10 MM / 10 KM

TEN THOUSAND TIMES SMALLER AND LARGER

At 10^{-4} meters, the thickness of a human hair, we're reaching the resolution limit of the human eye. If you look at alternating black and white lines any narrower than this, even under ideal illumination, what you see is gray.

Crystals of table salt—sodium chloride—vary in size quite a bit but the smaller ones are about this size.

Incredibly, there's insect life at this scale. The North American feather-winged beetle (*Nanosella fungi*, family *Ptiliidae*) is the smallest known species of beetle at just 0.25 mm long. Beetles of the genus *Goliatus* are 440 times as long as *N. fungi*, and thus approximately eighty-five million (= 440^{3}) times greater in volume. This is about the scale difference between a blue whale and a mouse, or between a dog and an ant.

N. fungi was once thought to be the smallest of all insects, but it has been dethroned by the blind, wingless parasitic wasp *Dicopomorpha echmepterygis*. Adult males are just 140 microns (0.14 mm). That's so small, they could easily fit inside large single-celled organisms such as some paramecia.

Parasitic wasps are a popular thing to be: *D. echmepterygis* is one of at least 100,000 species. They're so specialized that many are parasitic upon just one species of host insect.

The world's smallest insects—wingless parasitic wasps not even two tenths of a millimeter long—are to you as you are to Mount Everest or Mauna Loa.

Phytoplankton ('phyto' = plant; 'planktos' = made to wander) are single-celled marine algae. Some can move using little tails called flagella; others just drift with the water currents. These microscopic plant-like organisms float or swim in the upper 100 meters of the ocean, where they depend on sunlight for photosynthesis.

One type of algae you want to avoid is *Pseudonitzchia pungens*. Even under a microscope, this 125-micron nail-file looks almost identical to its harmless cousin *P. multiseries*; the slightly larger pores on *P. pungens* are your only warning that it produces domoic acid, the powerful neurotoxin that makes shellfish dangerous or even lethal to eat after 'red tides' or algal blooms. A microscope image blown up big enough to make it obvious which species is which makes the organism look the size of your finger. On the same scale, each of your eyelashes is the size of a tree.

Some phytoplankton, the algae known as diatoms, use silica (silicon dioxide, the main component of glass) to build their beautiful outer casings. There are about 10,000 species of diatoms. In warm seawater, you can scoop up 50,000 of them in a one-gallon jug.

"Peak XV" was measured at 29,002 feet by Andrew Waugh's Himalayan survey team of March 1856. Or rather, measured at 29,000 feet exactly: the extra two feet were added for fear that a round number would look like a mere estimate. Known as Sagarmatha ("goddess of the sky") to the Nepalese and as Chomolungma ("goddess mother of the world") to the Tibetans, the mountain was renamed after the surveyor Sir George Everest, who, to his credit, protested that the mountain should retain its native names.

Everest's height was revised upwards to 29,028 feet (8,848 meters) in 1955. The current figure, from a survey using GPS, is a rounder, more regâl one (at least in metric) of 8,850 meters (29,035 feet).

Whichever figure you choose, Everest has a metric height of over 8,000 meters—a distinction share by only 14 peaks in the world, all of them in either the Himalayas or the adjoining Karakoram.

The deepest point in all the world's oceans is Challenger Deep in the Mariana (or Marianas) Trench—look for a line of deep blue curling around the east and south of Guam; it's roughly 11,000 meters (36,000 feet) deep. If it swallowed Everest, even the wind-torn summit flags and discarded oxygen bottles (and, alas, about 200 bodies) would lie under more than a mile of seawater.

One of Earth's mountains is higher than Everest—sort of. Hawaii's Mauna Loa ("Long Mountain" in Hawaiian) has a summit "only" just over 4,000 meters above the sand and surf of Kona, half the height of Everest and slightly shorter than neighboring Mauna Kea ("White Mountain"). But most of Mauna Loa (another 4,500 meters, or 15,000 feet) is actually below sea level, so for a long time people have said that Mauna Loa is "really" as high as Everest. Recent studies reveal that in a sense Mauna Loa is even bigger: it's so massive (ten to twenty thousand cubic miles of rock, according to various estimates) that it has sunk into the surrounding sea bed. There's an additional huge chunk of the mountain—almost another 8,000 meters, in fact—buried in the seabed, for a total of nearly 17,000 meters—about ten miles. When I started writing this book, it was indisputably the bulkiest mountain on Earth ...

Jupiter's moon Io has several mountains as large as Everest or larger. Maat Mons, a volcano on Venus, peaks at over 11,000 meters. And Uranus' little moon Miranda has a line of *cliffs* that dwarf all these, at over 18,000 meters.

A neutron star is a much weirder object on this same scale. Stars up to 1.4 times the mass of the Sun shrink into white dwarfs. Stars larger than 3 times the mass of the Sun explode, and then become black holes. In between these limits, a supernova occurs but the star's remains collapse into the densest form of matter short of a black hole: a couple of solar masses trash-compacted by gravity into a ball the size of a small city. They are 'neutron' stars because their electrons have been crushed down into the atomic cores and have neutralized the protons: hence the entire star, to simplify a bit, is a super-dense ball of neutrons.

A single grain of "neutron star table sugar" would weigh 1,000 to 10,000 tons.

Scale Five: Cells and Craters

$10^{-5} / 10^{5}$

1/100 MM / 100 KM

A HUNDRED THOUSAND TIMES SMALLER AND
LARGER

We mentioned that the nematode worm *Caenorhabditis elegans* is a millimeter long. Its body is made up of about 1,000 cells, 300 of which are its brain. Most of these cells—like most of the cells in most living things—are about 10^{-5} mm.

Assuming you're a human and not a nematode worm, your brain has as many cells as there are stars in the Milky Way, and your body has hundreds of times more—about 10 to 100 trillion. But for every 'somatic' cell in your body (one that's genetically part of *you*) there are about ten usually much smaller hitchhiker cells such as bacteria and intestinal protozoa. You are a walking heap of over half a quadrillion (500,000,000,000,000) cells.

The single-cell eukaryotes are a staggeringly diverse group. They include both the protists or protozoae (single-celled "animals") and the protophyla (single-celled "plants"). Types of single-cell eukaryotes include dinoflagellates, amoebas, foraminifera, radiolaria, plasmodia (the cause of malaria, and thus possibly of more human deaths than any other agent), paramecia, diatoms and sporozoans. Notice that these aren't species names: they are broad *classes*: each of these classes in turn contains *Orders*, which contain *Families*, most of which contain more than one *Genus*, most of which contain more than one species. The known diatoms alone consist of 10,000 species divided into 250 genera.

A hundred of the beautiful, intricate, glasslike radiolara would fit onto the period at the end of this sentence.

Trypanosomes, which look like tiny pointed worms, are another class of single-celled creepies, and one of our worst enemies. Central Africa's tsetse fly delivers *Tripanosoma brucei* right into your blood; *T. brucei* gives you sleeping sickness. *T. cruzi* hitches a ride in an assassin bug (genus *Triatoma*), and jazzes up your life with Chagas disease. Some have theorized that a life-long cargo of *T. cruzi* was what gave Darwin such poor health.[3]

---

[3] DUTCH CURIOSITY: One of my heroes is Antony van Leeuwenhoek (pronounced LER-vun-huck). Born in Holland in 1632, Leeuwenhoek was a poorly-educated tradesman. But his insatiable thirst for discovery turned him into one of the most important scientists of all time. Using only his homemade magnifying lenses, he unlocked the door to an unexpected universe right under our noses: the place where bacteria, rotifers, and an endless zoo of other 'animalcules' live. He was fascinated by his own excreta, particularly by the way the 'animalcules' altered whenever he was 'troubled by a looseness.' He is summed up well in the brilliant understatement with which Josiah Wedgewood described his nephew, the young Charles Darwin—"a man of enlarged curiosity." When Leeuwenhoek's letters were published and translated, they caused amazement in the scientific establishment; in 1680, in a triumph of admiration over snobbery, he was elected to England's prestigious Royal Society.

Shrink to the size of a red blood cell and it will take you several days to hike across your breakfast table.

The Vredefort Crater, near Johannesburg in South Africa, was created 2 billion years ago by a 10 kilometer asteroid. It may be the largest and oldest impact site identified anywhere on Earth. The structure now visible is a 70 kilometer 'uplift dome' in the middle of the crater; the crater itself has sunk under other rock formations and disappeared, but was probably 300 kilometers across. That's about the size of Ireland.

The Chicxulib crater (centered on the village of Chicxulib, near Merida, on Mexico's Yucatan peninsula), may have been as big as Vredefort, but it's buried under thick layers of younger limestone, so its original size is hard to assess. It was created 65 million years ago when the Earth was hit by a meteorite 15 kilometers in diameter. The explosion incinerated everything within several hundred kilometers and created tsunamis in the Gulf of Mexico that washed *over* Florida into the Atlantic; meanwhile the real damage was done by molten rocks that were blasted into suborbital trajectories and landed pretty much everywhere, even on the other side of the planet, causing worldwide forest fires. The reduced sunlight and sudden global cooling certainly resulted in major climate change; it *may* have resulted in the great Cretaceous-Tertiary extinction, in which 50% of all land species followed the dinosaurs off-stage. (In the 1990s this looked like a certainty, but some evidence suggests Chicxulib may have happened 300,000 years too early to have fried the dinosaurs. The jury is still out.) A less famous but strikingly visible crater, linked to a much earlier mass extinction, is Lake Manicouagan in northern Quebec, Canada. The circular lake runs around a central uplift structure about about 70 kilometers across. Geologists think Manicouagan was formed by a meteorite near the end of the Triassic period about 212 million years ago—just at the right time to have caused the great Triassic extinction in which 60% of all species died out.

With keen eyes or a pair of binoculars you can see what Vredefort and Chicxulib looked like freshly made: just check out the full Moon. Copernicus and Tycho, the two largest impact craters on the Moon's nearside, are not quite as big as Vredefort and Chicxulib but, because the Moon is geologically inert and has no weather, they're almost as good as new—a billion years young, in fact—and because the Moon is smaller than Earth they are about the same relative size. Both are easily visible; at full Moon, you can also see "rays" radiating from Tycho that are lines of ejecta—material spewed out on impact.

People tend to think of the Hubble Space Telescope and the International Space Station as "in space" just the way the Apollo astronauts were "in space." But the Hubble and ISS typically orbit at a height of around 550 kilometers and 350 kilometers—they'd bump into two Long Islands stood on end. The Moon is *almost a thousand times* further away.

If the Earth is an orange, think of the Moon as a plum orbiting five paces away; the Hubble Space Telescope and the International Space Station are flecks of metal far too small to see that orbit three to five millimeters above the skin of the orange.

Until recently, we would have said that the largest of all known mountains dwarfs anything on Earth. But that's not true any more ...

Olympus Mons, on Mars, is a monster for sure. This mega-giant shield volcano sits on Mars' great Tharsis Dome, an upland area as large as North America, and shares it with at least three other volcanoes that make Everest look like a pimple: Mt. Ascraeus, Mt. Arsia and Mt. Pavonis. Olympus Mons is the champion: it rises 27 kilometers (17 miles) out of the flat surrounding plain, and is thus more than three times as high as Everest.

It's also 600 kilometers across—as wide as France—and it rises so gently that its peak is hidden from its base *by the Martian horizon*. Still, if you ever go peak-bagging on Mars you'll start your summit challenge with a fearsome climb. Most of the base is defined by a neat ring of cliffs, and the cliffs alone are at least 3,000 meters high. (Some astronomers give a maximum of 10,000 meters, which would make some of the cliffs higher than Everest.)

And now for a strange surprise: there's a mountain *on Earth* on almost the scale of Olympus Mons. Only discovered (or recognized for what it is) in 2013, the Tamu Massif lies under the Pacific Ocean west of Japan. It's nearly 4,500 meters high—shorter than Everest, but at 650 by 450 kilometers it covers an area about the size of the UK, nearly 300,000 square kilometers, or fifty times the base of Mauna Loa. You could pour Mauna Loa into it hundreds of times over.

Scale Six: Bacteria and Continents

$$10^{-6} / 10^6$$

A MICROMETER / 1,000 KM

A MILLION TIMES SMALLER AND LARGER

A single bacterium standing on your toe is like a single person standing on the beach at the "toe" of Italy.

A micrometer or micron (designated by the Greek letter μ, 'mu') is a thousandth of a millimeter. (It is to a millimeter as the millimeter is to a meter, or the meter to a kilometer.)

Deep red visible light, verging on infrared, has a wavelength of a bit less than half a micron; bright violet light, verging on ultraviolet, has a wavelength of less than three quarters of a micron. This tiny range gives us the entire world of color.

The difference between 0.4 micron light and 0.7 micron light is obvious in some of the largest things we can see. Betelgeuse, at Orion's right shoulder, is an old, cool star, and it shines mostly in the red end of the spectrum. Energetic young stars—like Vega, for example, directly overhead in the northern summer—are much hotter and therefore look blue or white. In fact, much of the "light" output of old stars falls out of view at the infrared end of the spectrum, and much of the light from very hot stars is equally invisible ultraviolet.

One micron is about the resolution limit of an optical microscope. Typical photosynthetic bacteria or cyanobacteria (also known as blue-green algae, although they aren't algae at all) are this size individually, as are the nuclei of many cells. The common bacterium of food-poisoning fame, *E. coli*, is slightly bigger, a rod-shaped structure 2 microns long and a micron or so in diameter. The common parasite *Cryptosporidium* is 3-5 microns.

Despite their size, bacteria are so abundant that they make up a huge portion—possibly 90%—of all ocean life by weight. Thousands live on every square inch of your skin; more like half a million per square inch in your armpit, and far more than that if you haven't washed recently. Your saliva contains 100 million per gram, a fertile gram of soil contains a billion (about as many as there are people in India), and a gram of your fecal material—or poop, as scientists prefer to call it—contains as many as a trillion; about 10 million of these are just the species *E. Coli*.

Change the scale so that a single bacterium on the end of your nose is the size of a pea. Then your nose is the size of a whale.

One of the smallest and most abundant classes of cyanobacteria, *Prochlorococcus*, was only discovered in 1988. Only about half a micron across (200 of them in a conga line would stretch the width of a human hair), these "picoplankton" float in seawater in such vast numbers—hundreds of thousands per milliliter, or the population of the United States in a bucketful—that they alone make up somewhere near half of all ocean life by weight.

Also on the micron scale are your chromosomes, the mostly X-shaped structures that carry your genetic information inside every cell.

The Caspian Sea, the world's largest body of fresh water, is

1,225 kilometers long. Siberia's Lake Baikal is only 650 kilometers long and 80 kilometers wide, but it's so deep (up to 1,600 meters) that it contains a fifth of the Earth's surface fresh water. But if the idea of a very deep lake like Baikal makes you think of a bathtub, you have the wrong idea. A scale model as long as a bathtub would be only five millimeters deep at its deepest point.

Asia is drained by dozens of rivers over 1,000 kilometers long and at least 20 over 2,000 kilometers long: these are the Amudarya (Oxus), Brahmaputra, Danube, Dnieper, Euphrates, Ganges, Indus, Irrawaddy, Kolyma, Lena, Mekong, Ob-Irtysh, Olensk, Salween, Syrdarya, Ural, Volga, Yangtze (Chiang Jiang), Yellow (Huang He), and Yenisey-Angara. Most of these rivers are longer than the Colorado and more than twice as long as the Rhine.

One impact feature on the moon is so big that it got named as a 'Mare' (sea) rather than the crater it is. Mare Orientale, only the eastern third of which is visible from Earth, has three distinct concentric impact rings and the outermost is 900 kilometers in diameter—almost twice as big as the Caspian Sea, or the same area as Madagascar. The largest asteroid, Ceres, is about the same size.

At 3,500 kilometers in diameter, the Moon itself is about the size of Australia; ditto Jupiter's moon Io.

Scale Seven: Viruses and Planets

$10^{-7} / 10^7$

1/10 MICROMETER / 10,000 KM

TEN MILLION TIMES SMALLER AND LARGER

Viruses are—by a different definition of "living thing"—the smallest living things of all. Living or not, they're really, really weird.

Viruses have no cellular structure at all; they're just chunks of DNA or RNA (genetic material), with a cloak of protein to hold them together. They have no metabolism and they do not eat, grow, or die: in many respects they are more like crystals than creatures. Yet they reproduce, essentially by injecting themselves into and taking over the cells of other living things.

And they are tiny. Even the biggest, the bacteriophages ("bacterium eaters") such as T4, are only two tenths of a micron long (200 nanometers, or 2×10^{-7} m), and half that wide.

Very small single-celled parasitic organisms, such as *Mycoplasma genitalium*, are as small as the T4 virus. Mycoplasms weigh a tenth of a trillionth of a gram. If every human on Earth donated a thousand of them, the resulting pile would barely smear the surface of a teaspoon.

Imagine 100 million people (the population of France and Spain combined) standing upright, shoulder to shoulder. If they were T4s, they could do this on the head of a pin.

Or imagine an ant the size of a blue whale. Its 'bacterio-phage' viruses would be the size of ants.

Even viruses may not be quite the end of the line. Stanley Prusiner discovered another, possibly smaller and more primitive class of infectious agents: "prions," or "proteina-ceous infectious particles." Prions are a type of protein, and unlike viruses they don't even carry any genetic material. Nevertheless, Prusiner showed that they can infect brain tissue and pass diseases like scrapie, kuru and Creutzfeldt-Jakob disease from one animal to another. A prion is a chain of a couple of hundred amino acids; there is some dispute about whether they are about the size of the smallest viruses or substantially smaller.[4]

The Andes of South America are over 7,000 kilometers long—more than the distance from the soles of your feet to the center of the Earth. But even they are dwarfed by a range that no mountain climber has ever explored.

The Mid-Atlantic Ridge, most of which is buried under a mile of seawater, describes an elongated "S" all the way from a point not far south of Greenland to Bouvet Island, which is just north of the Antarctic Circle. That's 15,000

---

[4] TOLD YOU SO: Stanley Prusiner began his research into 'proteinaceous infection' after treating patients with Creutzfeldt-Jakob disease in 1972; he published his results in 1982. Most of the scientific community laughed out loud at "prions": the idea of a single agent that could cause both communicable and heritable disease without having any genetic material was plainly ridiculous. But in 1997 Prusiner got the last laugh. Or rather, he got the one thing in science that's even better than the last laugh—a telephone call from Stockholm.

kilometers (9,300 statute miles or 8,000 nautical miles). Some say that even that is just one end of an essentially continuous chain that continues on to India, Australia and then Alaska.

The Nile and Amazon, placed end to end, would be slightly shorter: they're about 6,500 km each.

Asia is the Earth's largest landmass. Ignoring the artificial and purely cultural distinctions between Europe and Asia, and between India and Asia, it stretches more than 8,000 kilometers east to west, from Lisbon to eastern Siberia, and just a fraction less north to south, from Chelyuskin on the Aymir peninsula in northern Russia to Cape Cormorin at the southern tip of India.

The Pacific is the world's largest ocean, and also the deepest, both at its extreme depth and on average. It shrank suddenly in 2000, not because of thirsty whales but by command of the International Hydrographic Organization: all waters north of Antarctica, but south of the 60th degree line, are now defined as belonging to the newly-minted 'Southern Ocean.'

Still, the new, smaller Pacific is 18,600 kilometers diagonally across, from Taiwan to the coast of Chile. It covers 28% of the Earth's surface—roughly equal to all the land. (If you center a globe on a point in the south-central Pacific, say between Kiribati and Tuvalu, almost no land is visible.)

The human home, Earth, is 12,750 kilometers (just under 8,000 miles) in diameter at the equator. It weighs six billion trillion tons, or about a trillion tons per person. It has a surface area of just over 500 million sq kilometers (nearly 200 million square miles), or 70,000 square meters per person. Solar 'spicules,' gas jets rising from the Sun's surface, are often about this size. Jupiter's Great Red Spot is about as wide and three times as long.[5]

The Sahara desert is bigger than all the other major deserts in the world combined. But the Pacific is to the Sahara as a postcard is to a stamp.

Imagine you are as long as the Andes. The bacteria on your skin are like vast herds of elephants. Your nose alone rises to fifteen times the height of Everest; inside it, attached to hairs 20 kilometers long, are viruses the size of piglets.

---

[5] A SWIFTLY BULGING PLANET: A spinning object tends to bulge at the equator—it becomes an "oblate spheroid." Because Saturn is the least dense of all the planets, and spins very fast, it's more oblate than most: its equatorial radius is a tenth larger than its polar radius, and you can see this easily in photographs. But the Earth is rocky and dense, and rotates quite slowly, so it only bulges a little: its polar radius is just a third of 1% shorter than its equatorial radius.

Scale Eight: Small Bio-Machinery and Big Planets

$10^{-8} / 10^8$

1/00 MICROMETER / 10,000 KM

A HUNDRED MILLION TIMES SMALLER AND LARG-
ER

Ten billionths of a meter is the resolution limit of the Scanning Electron Microscope. It's a bridge between the living
and the molecular level. At this scale, we can see some of
the structures that very small organisms use to feed, breed
and move. Also visible here, like spaghetti in a bowl, is the
double-helix structure of DNA.[6]

The cell wall of a bacterium, a tiny fortress protecting a
lone strand of DNA, is about this thick.

We've already met the bruisers of the virus world, the
T4 and other bacteriophages. More typical viruses are less
than half as big. At the other end of the scale, some picornaviruses, a family containing human polivirus and rhi-
novirus (= "nose virus"), are as little as 14 nanometers.

---

[6] DNA'S SECRET HISTORY: Crick and Watson get all the
credit for discovering the double-helix structure of DNA. But
their work relied fundamentally on the discoveries of several
other people you've probably never heard of, most notably
Rosalind Franklin and Maurice Wilkins. (For more on the
same theme, see the notes on Edwin Hubble and on Eureka
Moments.)

If you plan to make jewelry out of picornaviruses, you will need 100,000 of them just to make a necklace long enough to go around the period at the end of this sentence.

Jupiter is 150,000 kilometers (1.5×10^8 meters) in diameter, more than 10 times as wide as the Earth. Jupiter's mass (equivalent to 318 Earths) makes it 2.5 times as heavy as all the other planets combined. But Jupiter would need to be almost 100 times more massive to become a star. Even brown dwarfs—20 to 80 Jupiter masses, but about the same size as Jupiter—are too small. A pity: an object with a tenth of the Sun's mass (100 Jupiter masses) will also be about the size of Jupiter, but it will burn. And red dwarfs sip their fuel so daintily that they can keep burning for a trillion years.

Scale Nine: The Buckeyball and the Sun

$$10^{-9} / 10^9$$

A NANOMETER / A MILLION KM

A BILLION TIMES SMALLER AND LARGER

The diameter of a DNA molecule is easily within range of modern microscopes at 2 millionths of a millimeter (2 nanometers). But this famously helical structure *itself* winds in a sort of helix, about 10 nanometer across. So you could argue that DNA should have been at Scale Eight.

Carbon's various crystal structures result in substances as radically different as graphite and diamond. In 1985, researchers working on long carbon chains accidentally created an entirely new form of the element, buckminsterfullerene (C_{60})—"Buckyballs." Named after designer and visionary Buckminster Fuller, C_{60} consists of pure carbon geoids a nanometer across. And (looking ahead to the next scale): since 2004 we have another wholly new form of carbon: graphene, a lattice structure just one atom thick.

The Sun is a medium-sized "main sequence" yellow star about 1.4×10^9 meters (a million miles) across. It's 12,000 times as far from us as the Earth is wide. It converts hydrogen to helium at the rate of 600 million tons per second. Fusion converts only 4 million tons of this mass into energy, and that's how much mass the Sun loses every second. But the Sun is so big that its nuclear fuel will last another 5 billion years, a million times as long as recorded human history so far. At the end of that time it will become a red giant, 100 times as big as it is now, and shortly afterwards will eject its outer layers and become visible from nearby stars as a planetary nebula, with a radius expanding to about twenty times the orbit of Pluto.

The Sun contains almost 99.9% of all the mass in the solar system. If the Solar System is a 500-sheet ream of copier paper, the Sun accounts for 499¼ sheets, Jupiter is half a sheet, and everything else—all the other planets, asteroids and comets—are the remaining quarter of a sheet.

Scale Ten: Atoms of Gold and Giants with Gas

$$10^{-10} / 10^{10}$$

1/10 NANOMETER / 10 MILLION KM

TEN BILLION TIMES SMALLER AND LARGER

We're down among the individual atoms now. The elements. Creation's little bricks.

The word 'atom' comes from the Greek for "not divisible." And we thought they were not, until the twentieth century.

Simple atoms such as hydrogen are about 10^{-10} meters (a tenth of a billionth of a meter) wide. The great American physicist Richard Feynman once said that, if he was trying to communicate with an alien, he might describe himself as "17 billion hydrogen atoms high."

But ask yourself this question: where do the elements *come from?* As we'll see when we answer that question, there's a romantic reason to pick the slightly larger gold atom (about 3×10^{-10} meters) as our standard.

There are 10^{17} (100 thousand trillion) atoms in a grain of sand. That's 14 million for you, 14 million for me, and so on, for every person on Earth.

Arcturus, an orange giant star in the constellation Boötes (*boo-OH-tease*), started life roughly the same size and type as the Sun, but it is several billion years older and has already converted most of its hydrogen into helium. Now it is becoming unstable as its nuclear furnace gobbles its way up the periodic table, turning helium into carbon and oxygen. At 26 times the width (3.5 x 10^{10} meters) it dwarfs our Sun.

Arcturus is to the Sun as the Sun is to the planet Uranus— and as Uranus is to Pluto. (Or: if Pluto is a sesame seed, then Uranus is a walnut, the Sun is a beach ball, and Arcturus is the size of a house.)

Arcturus is to you as you are to a gold atom.

Arcturus is easy to find. First find the Big Dipper (also known as the Plough, Ursa Major, and the Great Bear). Let your eye run to the left from the pan, along the three stars in the curved handle. Keep that arc in mind and continue it another two handle lengths. See a bright, orange star just above the point you reached? You have 'arced' to Arcturus. It's the fourth brightest star in the sky.

And now that romantic note. For most of human history, the origin of the elements has been a mystery. A metal like gold was simply found, and the question of where it came from was scarcely even asked. Perhaps large pressures created it in the Earth's geological past?

No.

By the 1920s it was understood that stars turn hydrogen into helium; by about 1940, it was known that carbon, nitrogen, and oxygen are involved, at least in larger stars. But it was not until the 1950s that the astronomer Fred Hoyle began to develop a full explanation for the origin of the elements, in the theory we now know as stellar nucleosynthesis. What the theory tells us is both beautiful and astonishing. The light elements (hydrogen, helium, and lithium) were created during the Big Bang at the beginning of the universe. More lithium, plus beryllium and boron, are made by cosmic rays in space. The middle elements like carbon, oxygen, iron, and nickel are cooked inside old stars. But the heavy elements can only be created in another, special way: they require a supernova, the cataclysmic death of a large star.

When large stars are turning hydrogen into helium, helium into carbon, and so on up through oxygen and silicon, these reactions all generate energy; the energy outflow supports (and, eventually, massively expands) the outer layers of the star. But silicon turns into iron—and iron *absorbs* energy when it is converted into heavier elements. So once the star's core is mainly iron, the outward energy flow is turned off, and the effect is rather like removing the pillars that hold up a roof.

The central nuclear fires very suddenly go out. Gravity makes the outer star collapse. All that mass has a long way to fall, and it accelerates to nearly the speed of light. The iron core collapses too, and the resulting nuclear interactions cause an explosion so violent that it can make a single star outshine, briefly, the rest of its galaxy. This is what creates the heavier elements, and scatters them like seeds through many light years of surrounding space.

Such stellar refuse, including minute quantities of gold, was part of the cosmic dust that congealed into our solar system. So every wedding ring is older than the Sun, a gift left to us in the will of a dying giant long before the Earth was formed.

Scale Eleven: Two Inner Solar Systems

$10^{-11} / 10^{11}$

1/100 NANOMETER / 100 MILLION KM

A HUNDRED BILLION TIMES
SMALLER AND LARGER

10^{-11} meters puts us well inside the atom's "inner solar system."

There's not much going on at this scale. It's tempting to say that we're inside the orbit of the electron, but such ordinary pictures break down here. To say that an electron is "in orbit around" the atomic nucleus proton is seriously misleading. Quantum mechanics tells us that at any moment the electron exists *with some probability at any point* within its "shell." It would be more accurate to say that the whole region is blurred or stained with "electron-ness."

Imagine your body as the lone proton at the center of a hydrogen atom. On the same scale, the electron's little kingdom is 100 km across.

Earth orbits the Sun at a distance of almost exactly 1.5×10^{11} meters (150 million kilometers, or 93 million miles)—one "astronomical unit," or AU.

Scale Twelve: Light Waves and Light Hours

10^{-12} / 10^{12}

A PICOMETER / A BILLION KM

ONE TRILLION TIMES
SMALLER AND LARGER

10^{-12} meters—a trillionth of a meter, or one picometer—is the wavelength of a gamma ray. There's not much going on at this rung of the microcosmic ladder either, so it's a good place to look at some other wavelengths.

As explained in Appendix 3, the 'speed of light' (often cited in relation to Einstein's theories) is really just the speed of all electromagnetic energy, light included. But different types of energy have different 'energy levels' associated with them, and their wavelengths reflect this: the shorter the wavelength, the more energetic (and potentially dangerous) the radiation.

Wavelengths in the electromagnetic spectrum vary continuously from the size of a county to the size of the atomic nucleus. Some examples:

- ULF ("Ultra Low Frequency") radio waves: several kilometers
- Low frequency radio waves: a kilometer
- AM radio: several hundred meters
- TV and FM radio: 1-10 meters
- Microwaves: a centimeter

- Infrared light: about a micrometer (a millionth of a meter)
- Visible light: four to seven tenths of a micrometer (a bit less than a millionth of a meter)
- Ultraviolet light: less than half a millionth of a meter to 10 billionths of a meter
- X-rays: around a nanometer (a billionth of a meter)
- Gamma rays: around ten picometers (trillionths of a meter)—less than the width of even a small atom

One light hour is 10^{12} meters, or about 7 AU. That's twice the diameter of the orbit of Mars—a comfortable envelope for the four rocky planets and asteroid belt that make up the inner solar system. It's also the diameter of a red supergiant star like Antares or Betelgeuse.

If the Sun is a sesame seed, Arcturus is a small marble and Antares is the size of your head.

The hypergiant star VY Canis Majoris is twice the diameter even of Antares. One of the largest stars known, it would fill our solar system to well beyond the orbit of Saturn. You would need five thousand Suns to make a necklace around its equator—or half a million Earths. To get a sense of the surface area, consider a *billion* Pacific oceans: quilted together, and thrown down onto the surface of VY Canis Majoris, they would take up one percent of one per cent of the available space—the same as Northern Ireland on Earth.

If Earth is a grain of salt, and the Sun is a mouse, VY Canis Majoris is twice the size of a blue whale.

Even when a thousand times bigger than the Sun, most hypergiants are only 25-50 times greater in mass, with a rare few at 100+ solar masses, and even fewer up to around 200+ solar masses. But stars this big are immensely unstable and burn through their nuclear fuel thousands of times faster than medium-sized stars. Having already converted all their hydrogen into helium and lithium, they are turning those elements into carbon and oxygen: they are close to collapse and supernova.

The hypergiant Rho Cassiopaeiae is so unstable that astronomers think it could 'go supernova' at any time. In fact, for all we know it exploded in the time of the Pharaohs: it's 10,000 light years away, so the announcement may still be in the mail.

Scale Thirteen: A Desert and the Solar System

$10^{-13} / 10^{13}$

1/10 PICOMETER / 10 BILLION KM

TEN TRILLION TIMES SMALLER AND LARGER

The 10^{-13} meter scale (a tenth of a trillionth of a meter) is one of nature's Empty Quarters. At this scale you're already inside the inner electron shells of an atom, yet the atomic nucleus is so much smaller that it's still a tiny, distant object.

The ring formed by Neptune's orbit is about 10^{13} m—a good round number for the size of our planetary system.

Over 10^{13} meters (74 AU) is the *closest* approach to the Sun of Sedna, the most distant known dwarf planet and the one with by far the most eccentric orbit—see Scale Fourteen.

If you plan to be around in 2061, mark your calendar for the return of Edmund Halley's comet, which he saw in 1682 and correctly predicted would return about every 76 years. Comet Halley has an elongated orbit that brings it well inside Earth's orbit, and then takes it 35 AU away, out beyond Neptune. First recorded thousands of years ago, and immortalized in the Bayeux Tapestry (depicting the Norman invasion of England in 1066), it's a peanut-shaped ball of ice and rock about twice the size of Mount Everest.

Scale Fourteen: The Heart of the Atom and the Beginning of the Interstellar Desert

$$10^{-14} / 10^{14}$$

1/100 PICOMETER / 100 BILLION KM

A HUNDRED TRILLION TIMES SMALLER AND LARGER

The nucleus of the gold atom—79 protons and 117 neutrons—is 1.4×10^{-14} meters across.

1.4×10^{14} meters is nearly 1,000 AU—the most distant point from the Sun of Sedna's wildly elliptical 10,000 year orbit. That's enough to take us well out of the familiar planetary system, well beyond the Kuiper Belt, and near to the beginning of the very diffuse Oort Cloud, the home of the long-period comets.

A gold nucleus is to you as you are to the far outer solar system.

Get a piece of chalk and draw a long thin ellipse on the ground that's just big enough to lie down in. A hand's width inside one end, make a dot two millimeters in diameter. If the ellipse is the orbit of Sedna, the dot is the orbit of Earth.

Scale Fifteen: Protons and the Interstellar Medium

10^{-15} / 10^{15}

A FEMTOMETER / A TRILLION KM

A THOUSAND TRILLION TIMES SMALLER AND LARGER

You could say that the heart of the atom, the proton, is about 2×10^{-15} meters wide.

But something odd is going on here. Its size was something physicists thought they had down exactly: a radius of 0.8768×10^{-15} meters. But new measurements (2014) suggest it may actually be (are you sitting down?) 0.84087×10^{-15} meters.

OK, this sounds like a comical difference. But look closer: it's like discovering that all our modern, triple-tested, super-careful, GPS-and-laser-aided measurements of Mount Everest, down to the nearest few inches, have been out by a thousand feet.

10^{15} meters, or a trillion kilometers, is a bit more than a twentieth of a light year, and there's not much more going on at this scale than at the last. A twentieth of a light year is much bigger than the solar system, if you take the orbit of Neptune as the boundary, and much smaller than the solar system if you take the outer Oort Cloud (Scale Sixteen) as the boundary. It's the edge of interstellar space.

But there are some things about this size. The strange object known as 3C48, discovered in 1960, looked like a star in some ways but was doing crazy stuff for a star—hence *quasi-stellar object* or quasar: 3C48 was the first one ever detected. It is probably about a tenth of a light year across, yet it radiates as much energy as a galaxy. Luckily for us, it's four billion light years away.

What's in interstellar space? A vacuum, more or less. But all space—even intergalactic space; even the vast bubble-shaped voids between the galaxy superclusters—has one thing in it, a faint energy at the frigid temperature of 2.7 degrees above absolute zero: the Cosmic Microwave Background (CMB) radiation.

The CMB is nothing less than the dying roar of the Big Bang. Predicted by George Gamow in 1948, it was discovered more or less accidentally in 1964 by Arno Penzias and Robert Wilson, who famously thought, at first, that the annoying hiss they kept hearing in their New Jersey microwave antenna was interference caused by "white dielectric material"—pigeon poop. In fact they were 'hearing' the half billion photons per cubic meter with which the CMB has littered all of space, including the cubic meter that's currently parked in front of your nose.

AND HERE WE ENCOUNTER A PROBLEM ...

There are amazing structures ten times bigger than those on the previous page—and there are other structures ten, a thousand, a million times bigger even than this.

But the very small is different.

Until recently, scientists believed that infinitely small sizes made physical sense, and that the scale of the atomic nucleus, at around 10^{-15} meters, was as far as our understanding was likely to extend any time in the foreseeable future.

Then, in the 1920s, quantum mechanics put an absolute limit on the very small, and, in the decade or so after 1980, physicists proposed the existence of a phenomenon at that absolutely smallest scale. They called it string.

Superstring theory, or just string theory for short, says that the smallest components of matter are not particles but vibrating lines or loops. And these pieces of "string" are ... small.

Imagine you live in a world without large animals. You and your tribe are familiar with mice, but there's no animal life bigger. Then one day you go hunting, and you come back to the fireside with bizarre news.

"Over those hills I've found a new kind of creature. They're ... they're ... big!"

Your friends ask how big these new creatures are. "Even bigger than mice?"

"Well," you say, not sure they will believe you. "Yes— bigger than mice."

"Really? How much bigger?"

"On average, about, uh, ten times the size of the solar system."

String theory is like that.

String theory says nothing about 10^{-18}, or 10^{-19} or 10^{-20}. It leaves these and many more orders of magnitude emptier than the most wholly uninhabited desert, and locates the next interesting physical structure at a scale that is, even by the Lilliputian standards of the atomic nucleus, extravagantly tiny: 10^{-35} meters.

We'll see later on just how mind-boggling a scale this is. (Briefly: it's so mind-boggling that saying "it's mind-boggling" leaves your mind not even remotely boggled enough.) For now, all we need to note is that the next thing in the physics of the very small is going on so far down the microscope that we are left with a gap many orders of magnitude wide. As we continue our journey up the astronomical ladder, there is simply nothing to say about this realm of the very small.

That's why the next few pages deal only with the ultra-large. We will meet strings again at the very end.

Scale Sixteen: The Solar Exurbs
and the Corpses of Stars

10^{16}

A LIGHT YEAR

TEN THOUSAND TRILLION TIMES LARGER

The Oort Cloud, a great diffuse shell of trillions of "long-period" comets, is the "outer outer" solar system, about a light year from the Sun. Comets from out here revisit the inner solar system regularly, just like Halley's Comet, but only once every few thousand years.

Put a large coin on a small round table. If Earth's orbit is the edge of the coin, and the Kuiper Belt is the edge of the table, the Oort Cloud begins several miles away.

From a vantage point within the Oort Cloud, you are still within the gravitational reach of the Sun, but it now so far away that it looks like any other bright star.

'Planetary nebulae' have nothing to do with planets, except that some of them look misleadingly like planets: they are what's created when middling-sized stars die. Main sequence stars (such as the Sun) turn into red giants as they run out of hydrogen; through processes that are still not well understood, they then eject 10-20% of their mass as a gas shell. The central remnant is a white dwarf star only the same diameter as the Earth, but the gas shell we observe is often about a light year across—about the diameter of the

Oort Cloud. Both NGC 5307 in Centaurus and M57 (the Ring Nebula, in the constellation Lyra), are on this scale.

Up towards the end of this scale we start to meet other stars. Alpha Centauri is four or five times as far away as the Oort Cloud, and 300,000 times as far away as the Sun. (It's also known as *Rigel Kentaurus* , the Centaur's foot—not to be confused with the star Rigel, which is Orion's foot). Alpha Centauri is often said to be the nearest star, but this is wrong in two ways. First: "Alpha Centauri" is really the twin-star system Alpha Centauri A and Alpha Centauri B; both stars are about 4.4 light years away and both are about the same size as the Sun. (Taken together, they form the third brightest star in the sky after Sirius and Canopus.) Secondly, they are only the nearest stars visible with the naked eye: the nearest star of all is their much smaller red dwarf companion, Proxima Centauri, 4.2 light years away.

The scale and emptiness of interstellar space is hard to understand, and genuinely appalling when you do understand it. The Alpha Centauri system, our nearest stellar neighbor, is *seven thousand* times as far away as Pluto.[7]

If Alpha Centauri A and B are two fleas circling each other a few paces apart, the Sun is a third flea twenty miles away. The star Betelgeuse, on this scale, is a large orange dog three thousand miles away.

---

[7] **LAST GAS FOR A LONG, LONG TIME:** The difference between 4.4 light years (Alpha Centauri A and B) and 4.2 light years (Proxima Centauri) doesn't sound like much, but it's over a trillion miles. If you tried to road trip from Alpha Centauri A to Proxima Centauri—in a very fast car, putting in eight hours a day at the wheel with not too many pee breaks —the drive would take three million years.

Scale Seventeen: The Stellar Neighborhood

10^{17}

TEN LIGHT YEARS

A HUNDRED THOUSAND TRILLION TIMES LARGER

10^{17} meters is the height of the 'Pillars of Creation,' those tall columns of dust and gas made famous by the Hubble telescope. It's also the scale of our local stellar neighborhood.

One famous nebula, the Crab (also known as Messier 1, in Taurus), is 10 light years across. At least 5,000 light years away, it is the remains of a giant star that exploded. Chinese astronomers noted the supernova in July 1054, which means it actually exploded around 4,000 B.C. (Earth time)—it's just that the news took 5,000 years to arrive. It's still expanding in all directions at over 1,000 kilometers a second. In the middle, and giving off powerful radio pulses thirty times a second, is a rapidly spinning neutron star—the remnant of the cataclysm.

The Orion nebula, near Orion's belt, is bigger still at 24 light years across and, at 1,300 light years away, is the closest significant star-forming region to Earth.

The "dog star" Sirius, yapping at the heels of Orion the hunter and the brightest star in the sky, is a little less than 10 light years distant. You are within 20 light years of about 100 stars.

Here are some neighborhood stars, with comparisons to the Sun:

- **Alpha Centauri A** and **B**, aka Rigel Kentaurus, 4.4 light years away, and both very roughly the size of the Sun

- **Sirius A**, 8.6 light years away, twice the size and 20 times as bright

- **Sirius B**, a white dwarf companion to Sirius A, the same mass as the Sun but smaller than Earth

- **Epsilon Eridani** and **Tau Ceti**, both about 10-12 light years away, three-quarters the size, and less than half as bright

- **Vega**, the bright blue star in the northern "summer triangle" that featured in Carl Sagan's novel *Contact*, 24 light years away, about twice the mass and 3 times the size, but 50 times as bright

- **Arcturus**, 37 light years away, 25 times the size, and 200 times as bright.

Scale Eighteen: Globular Clusters

10^{18}

A HUNDRED LIGHT YEARS

A MILLION TRILLION TIMES LARGER

Throughout the Milky Way there are "globular clusters" of stars, gravitationally-bound star-cities with 100,000 up to millions of members. Of those visible from Earth, the most beautiful is Omega Centauri, in the southern constellation Centaurus. It's about 100 light years in diameter—though estimates vary wildly—and it has about 10 million stars.

If our solar system were the size of your fist, Omega Centauri would be the size of Mount Everest. If the Sun were the size of your fist, Omega Centauri would be the size of Jupiter.

Scale Nineteen: A Distant Horse

10^{19}

A THOUSAND LIGHT YEARS

TEN MILLION TRILLION TIMES LARGER

Rigel—*Rigel Orionis*, Orion's left foot—is about 800-900 light years away. You can still see it clearly because it's 60 times the diameter of the Sun and about *60,000* times as bright.

1,500 light years is the distance to one of the sky's most hauntingly beautiful objects, the Horsehead Nebula. A dark swirl of gas and dust, it resides in the constellation Orion, just to the right of the star Zeta Orionis, and is illuminated from behind by the red light of another, separate emission nebula. The head itself is about 25 light years high; equestrians may be interested to note that this implies a complete horse of four trillion trillion hands. Feed it some sugar lumps, if you can: you'll need the "half a cubic light year" size.

Our solar system would scarcely make a visible dot against this horse's black head; even if we stretch a point, and imagine the solar system as a luminous disk all the way out to the fringes of the Oort Cloud, it would only be big enough to give the Horse a nosebag.

If California were a thousand light years long, the bacteria in a movie star's mouth would be like clusters of planets.

The 1,500 light years from Earth to the Horsehead Nebula is a bit more than 1% of the distance from one end of the galaxy to the other. We lie close to the edge of the galaxy in the Orion arm; the Horsehead is more than three times farther away from us than Betelgeuse in the direction of the galactic edge but, in galactic terms, that makes it just a hair farther out. The sky behind them is relatively dark because there's nothing much 'behind' them except intergalactic space.

Scale Twenty: Magellan's Clouds

10^{20}

TEN THOUSAND LIGHT YEARS

A HUNDRED MILLION TRILLION TIMES LARGER

The Orion Arm of the Milky Way, in which we live, is about 10,000 light years long—or 600 million times the distance from here to the Sun.

Two of the most luminous stars known, the hypergiants Eta Carinae and Rho Cassiopeiae, are about this far away. Both are several hundred times the diameter of the Sun; by some estimates, Rho Cassiopeiae is five million times as bright.

The nearest galaxies to us are the Large and Small Magellanic Clouds—irregular satellite galaxies of the Milky Way. They're easily visible from the Southern hemisphere as blurs of light. The Large Magellanic Cloud is 10,000 light years across.

If a whale were the size of the Orion Arm, the individual droplets of water vapor in its spout would be a thousand times larger than the largest stars.

Scale Twenty-One: The Milky Way

10^{21}

A HUNDRED THOUSAND LIGHT YEARS

A BILLION TRILLION TIMES LARGER

The Milky Way is a fairly typical spiral galaxy containing about two hundred billion stars. Seen from the outside it's a flat, thin disk with a central bulge—a fried egg 100,000 light years across, two or three thousand light years thick at the edge and 10,000 light years thick at the yolk.

But the "flat spiral" of popular illustration is misleading. The globular clusters of mostly older stars mainly occupy a spherical 'halo' about the same diameter as the denser and more obvious disk. Some of these 'halo' objects are very distant, forming a scarcely-detectable outer suburb of the galaxy rather like the Oort Cloud in the solar system. Some globular clusters are as much as 400,000 light years from the galactic center. That's a fifth of the way to Andromeda.

A convenient model of the solar system uses a "shrinkage factor" of 6 billion, making it a very walkable two kilometers wide. But that's not enough for getting a grip on the Milky Way. A model of our *galaxy* at that scale would barely fit inside Earth's orbit. You could hike around in it for the rest of your life and get almost nowhere.

So what happens if we shrink everything by a few more orders of magnitude? Instead of dividing by 6 billion, we'll try 600 million billion; that gets us a spiral 1.5 km in diameter—a model you can fit into a large city park. A model

Milky Way 1.5 km across is easy to walk around in. Unfortunately it's also invisible: everything it contains is too small to see.

Even hypergiant stars on this scale are the size of bacteria. A star like the Sun is no wider than a DNA molecule—beyond the reach of most microscopes. Large gas planets like Jupiter are the size of atoms.

Still, if we cheat and turn up the wattage a little, we can imagine strolling through a tenuous mist of light and pick out some of the main features. The central bulge is 150 meters thick: a steep but modest hill and its mirror image. Our solar system is (like the largest stars) the size of a bacterium.

The Alpha Centauri system is a hand's width away. Betelgeuse is seven or eight paces away. The Horsehead Nebula is another 15-20 paces further in the same direction.

Outside the Milky Way, the nearest objects are the Large and Small Magellanic clouds, fuzzy blobs a mile or two away. The Andromeda galaxy is a hazy swirl like the one we're in, twenty miles distant.

On this micro-miniaturized scale, the Hydra-Centaurus-Pavo Galaxy Supercluster is still 2,000 miles away.

Imagine the disk of the Milky Way as a small round café table. Our solar system is an invisible speck not far from the edge, and it orbits the galactic center just as Earth orbits the Sun. Since the invention of writing, around 3500 BCE in ancient Sumer, it has moved a tenth of a millimeter.

Scale Twenty-Two: The Local Group

10^{22}

A MILLION LIGHT YEARS

TEN BILLION TRILLION TIMES LARGER

Until 1924, most people thought the Milky Way was the whole universe. In that year, the great Edwin Hubble came down from the 100-inch telescope atop Mount Wilson with epoch-making news: at least some stars in the Andromeda "nebula" were much, much too far away to be part of the Milky Way.

Hubble had finally proven eighteenth-century philosopher Immanuel Kant's "island universe" theory—many of the "nebulae" were independent galaxies, not objects inside the Milky Way.

If you walked backwards into space for about a million light years, say in the direction of the Pole Star or the Southern Cross, you would see the Milky Way as a complete spiral—and you would see another, similar spiral nearby. The Andromeda 'nebula,' or galaxy as we now say, is just over 2 million light years away, and is often given as the most distant object visible with the naked eye.

Midway between Andromeda and the Milky Way is the gravitational center of the galactic clump known as the Local Group, which also contains about forty smaller galaxies. The Local Group fits into a sphere of space about 4 million light years across.

Imagine a very large, empty warehouse. **Suspended from the ceiling at varying heights are two one-meter-wide truck tires, a couple of dozen basketballs, and a couple of dozen softballs. The two truck tires are the Milky Way and Andromeda; the basketballs and softballs are the minor galaxies.**[8]

---

[8] ANDROMEDA ... MOST DISTANT OBJECT: Pick a clear, dark night and give your eyes twenty minutes to adjust. Then look for the 'broken W' of Cassiopeia, and look down from the W and to the right until you see the great square of Pegasus: Andromeda is halfway between them. It's easy to miss, but worth the trouble: the photons from Andromeda that will fall into your eye have been traveling towards you since a billion years before Earth's first multicellular organisms evolved.

Under exceptional 'seeing' you might also pick out a smaller, more distant spiral in the Local Group, Messier 33 in the constellation Triangulum. M33 is 2.4 million light years away. There are even some claims by exceptional observers to have 'eyeballed' galaxies such as Messier 81 at five times that distance. But even Andromeda is tough to spot.

Scale Twenty-Three: Clusters and Monsters

$$10^{23}$$

TEN MILLION LIGHT YEARS

A HUNDRED BILLION TRILLION TIMES LARGER

We shouldn't be too quick to think of galaxies as "small" relative to "large" structures like the Local Group. The largest individual galaxies aren't spirals like Andromeda and the Milky Way: they are "cD class" giant ellipticals. IC 1101 is a monster at the heart of the galaxy cluster Abell 2029 in Virgo. It may have a diameter of 8 million light years, in which case it has eight times the volume of the entire Local Group. Surrounding it is a bee-swarm of 5,000 to 15,000 globular star clusters. The whole thing may contain fifty trillion solar masses.

If the Milky Way is Boston, and Andromeda is New York City, the giant elliptical galaxy IC 1101 stretches from New England to the Carolinas—and our entire solar system out to Neptune is a grain of sugar in Harvard Square.

Scale Twenty-Four: Walls and Foam

$$10^{24}$$

A HUNDRED MILLION LIGHT YEARS

A TRILLION TRILLION TIMES LARGER

The Coma Supercluster, 320 million light years away, is sheet-like aggregation of galaxies so big that it barely belongs on this scale. Better known as the Great Wall, it's five hundred million light years long: when it was discovered in the 1980s, it was the largest known structure in the universe. Look "under" the Big Dipper—which you can find on any clear night in the northern hemisphere because it's "circumpolar" (it never sets below the horizon). Now find the constellations Hercules, Boötes, and Leo. The Great Wall is a million times farther away than those stars, but it stretches right across the space behind them, straddling a quarter of the sky.

Think of the Great Wall as a bent mattress six feet long, a few feet wide, and six inches thick. On this scale, the almost unimaginable gulf dividing the Milky Way from Andromeda (2 million light years) is the width of your thumb.

Structures like the Great Wall are typically curved skeins or filaments surrounding giant voids, some of which are areas several hundred million light years across and contain almost no galaxies at all.

Scale Twenty-Five: Quasar Country

$$10^{25}$$

A BILLION LIGHT YEARS

TEN TRILLION TRILLION TIMES LARGER

The Sloan Digital Sky Survey, begun in 1998, discovered a structure even more monstrous than the Coma Supercluster. The "Sloan Great Wall" is a billion light years away and almost 1.5 billion light years across.

Imagine walking across Texas, from Houston to El Paso—it will take you a month or two. At some point you kneel down, put your face close to the ground, and notice one very small grain of dirt, just barely big enough to see. Our own solar system bears the same relation to the Sloan Great Wall as that grain of dirt does to Texas.

When radio astronomers first observed quasars, in the 1950s, they just didn't make any sense. These incredibly bright and powerful "quasi-stellar radio sources" were star-like, being very concentrated. In fact, astronomers were so convinced that they must be stars, even though all sorts of facts about them didn't make sense, that it took several years for anyone to stumble on the truth.

It turns out that the typical quasar, say the object known as 3C273, is not a strange kind of star a thousand light years away. It's a strange kind of galaxy, receding from us at perhaps 80% of the speed of light, and already at least a billion light years away. Most quasars are much further away even than that.

Every galaxy probably has a massive black hole at its center, and our best current guess is that quasars are an early and very violent stage in the creation of these galactic black holes.

Scale Twenty-Six: Everything?

$$10^{26}$$

TEN BILLION LIGHT YEARS

A HUNDRED TRILLION TRILLION TIMES LARGER

We're not done with walls yet: the 'Hercules-Corona-Borealis Great Wall,' discovered in 2013, is currently the largest known structure in the universe at *10 billion light years by 7 billion light years*—twenty times as long as the original 'Coma Great Wall.'

People tend to confuse "the universe" with "the *observable* universe." As we'll see, this may be a truly enormous mistake. But the size of the observable part (also known as our "Hubble volume") is something we are starting to understand with some precision. Two generations ago, there was no general agreement about the Big Bang even happening. A generation ago, people were saying it happened "ten to twenty" billion years ago. Nearly everyone now agrees that our universe expanded from an almost infinitely dense point *thirteen-point-seven* billion years ago.

OK—just suppose for the sake of argument that the exact figure is 13.70038 billion years. You'll see the reason for this funny-looking number in a minute.

Light has to lope along at c (about 300 million meters per second) to reach our telescopes. So if Galaxy X is 10 billion light years distant now, we see it as it was 10 billion years ago: the photons coming down our telescope are the ones it emitted *then*, not the ones it's emitting now.

But photons themselves were unable to move around in space freely until 380,000 years AB (After Bang). Up to then, the universe was a hot plasma, a soup of free protons and electrons, in which stable atoms could not form. Photons (light) kept bumping into the free electrons, and could not move for any distance in a straight line.

Soup cools if you spread it out. So does a universe. Eventually, the primordial soup cooled to about 3,000 degrees Centigrade, at which point the positively-charged protons could combine with neutrons and with all those negatively-charged electrons. Bingo: an entire universe of helium (about a quarter of the atoms) and hydrogen (the other three quarters) was formed—and suddenly the photons were free to push through the crowd.

This "photon decoupling," rather than the beginning of time, marks the moment when light first streamed freely across space. Which has the curious implication that you would not have witnessed the Big Bang as a fireball: for a third of a million years, the mother of all explosions progressed in total darkness.

We can still detect the light that emerged when decoupling occurred—sort of. Because space itself has been expanding, that "first light" has literally stretched until it is no longer in the visible part of the spectrum. Its original wavelength was less than a micron, but now it has been stretched (by a factor of a thousand) into millimeter microwaves. Yes, reheat that galaxy for lunch: it's the Cosmic Microwave Background!

Now back to "13.70038." Since we cannot, even in principal, "see back" beyond photon decoupling, our most distant observable object must be at most 13.7 billion light years away (= 13.70038 billion light years minus 380,000 light years). And thus our 'observable universe' is at most

13.7 billion light years in radius, or 27.4 billion light years in diameter. That's about 3×10^{26} meters—a quarter of a million Milky Ways laid end-to-end.

Actually, thanks to Einstein, it's a bit more complicated than that. If light came from a galaxy 14 billion years ago, then that galaxy isn't 14 billion light years away now. It was 14 billion light years away when the light was emitted—but the universe has continued to expand since then. Think of a friend standing 20 feet away and throwing you a ball. But your friend is standing on a rubber sheet that's being stretched as the ball flies through the air. By the time you catch the ball, she's (say) 30 feet away, not 20.

In the same way, an object that threw its light at us 14 billion years ago is now more like 45 billion light years away. So our 'lookback time' is about 14 billion years (that's the oldest light we see) but the objects that emitted that light must now be much further away. Hence the 'real' diameter of the observable universe (the space now occupied by all the objects we can see) is at least 90 billion light years, or almost 10^{27} meters—another whole order of magnitude 'up.'

But this number, the size of the *observable* universe, has *nothing to do with* how much universe is "really out there." The edge of the observable universe is an edge in time, not space. If there are galaxies further than about 13.7 billion light years away, we simply can't see them, since the universe hasn't been around long enough for their light to reach us.

Scale Thirty-Five: The Kingdom of Planck and String

$$10^{35} / 10^{-35}$$

A HUNDRED BILLION LIGHT YEARS

A HUNDRED BILLION TRILLION TRILLION TIMES SMALLER

Max Planck probably doesn't deserve to be as famous as Einstein, but he deserves to be much more famous than he is. Among many other things, in 1900 he discovered the crack in the dam of Victorian physics that made both quantum mechanics and much of Einstein's work possible.

Matter was supposed to come in chunks; energy was supposed to be wavelike and infinitely variable. But in 1900 Planck discovered that energy is chunky too: it comes in multiples of a basic unit or quantum. This "quantum principle," one of the most profound discoveries in the history of thought, is one of the main (forgive me) planks on which all physics now rests.

One implication of Planck's discovery is that there ought to be a minimum quantum of space, or length. "10^{-35} m" gives you the idea, though, to be precise as possible for once, it's really $1.61619926 \times 10^{-35}$ meters.

The Planck length is to the period at the end of this sentence as the period at the end of this sentence is to the observable universe.

As I said, string theory envisages each "point" in our space as really a vibrating "string" about one Planck length across. Remember that a proton is 10^{-15} meters across: that means you need 10^{20} of them, side by side in a line, just to stretch across a proton.

Make an effort to get a grip on "10^{20}" here. (You'll fail. Try anyway.)

People standing side by side in a line, holding hands like figures in a paper chain, take up about a meter each. A line of 10^{20} such people would stretch from your front door to the star Sirius. And back. Five hundred times.

Expand a superstring to the size of a proton; then a proton is bigger than New York City. Expand a superstring to the size of a barely visible speck of dust; then the speck of dust is bigger than the observable universe. Expand a superstring to your size; then each atom in your body is almost as big as the observable universe.

Many versions of string theory have emerged—too many: 10^{500} of them, by some calculations. And many people think weird 1-dimensional strings are just a special case of even weirder many-dimensional objects. Since a "membrane" is a 2-dimensional surface, strings are sometimes referred to as "1-branes," and a three-dimensional space wrapped around a 4-dimensional space would then be a "3-brane." Because you can't do cutting-edge science without a bad sense of humor, these objects are known collectively as "p-branes."

Scale !?!: The Inflationary Multiverse

$$10^{1,000,000,000,000} \text{ ?}$$

MORE-WORDS-THAN-YOU-COULD-FIT-ON-THE-
PAGE BIGGER

In the late 1970s, Alan Guth of M.I.T. was puzzling over some unsolved problems in cosmology. One day, a theory about the early universe popped into his head that was interesting because of the neat way it seemed to solve some of those problems. The theory had some truly *ludicrous* consequences, but it was still kind of interesting. Guth called his brainchild *inflation*.

Theorists will be theorists. No need to worry, right?

Wrong.

A generation later, although there are a few dissenters working on alternatives, the majority of cosmologists now think that some version of inflation is probably, to everyone's great surprise, *true*. Because of this, you are about to get a very bad headache.

You may be thinking: 10^{27} meters! That's a hundred billion light years! Which is, like, almost as big as next week! How much worse can it get?

The answer is: worse.

Worse than worse.

Far more worse than you can possibly wrap your 15 billion neurons around.

But a little 'back story' is important here, because this is the point at which—in one of the most profound and hauntingly beautiful conceptual shifts in the history of science—the minuscule world of subatomic physics and the vast world of cosmology join hands.

Physicists like Alan Guth were looking for GUTs, "grand unified theories" that would combine general relativity (Einstein's theory of gravitation) with quantum mechanics (the theory of everything else), thus making sense of all the forces of nature in one framework. (I'll skip the suspense: more than a generation later they're still sitting on the pot, purple in the face, grunting and straining. Nothing but gas has yet been produced.)

A key question for grand unification theories is: what happens to particles at higher and higher energies?

That question is why they keep building really huge, really expensive machines. The biggest and most expensive (17 miles around, over $5 billion to build, and an annual electricity bill of more than $20 million), is CERN's Large Hadron Collider. When fully completed, it will be able to boot a single particle to an energy level of 7 GeV (Giga-electron volts). But Guth wanted to know what happens when a single particle has 10^{14} GeV—trillions of times more.

You could investigate this, in theory, by building a new particle accelerator larger than the solar system. If funds don't stretch that far, the only available 'lab' is the first 10^{-35} of a second after the Big Bang. At that time, particles *typically* had energies of 10^{14} GeV apiece.

That's how Alan Guth came to be one of a group of young physicists wishing they knew a bit more cosmology. And that's how he came to be familiar with all sorts of problems about the 'standard' Big Bang theory that the cosmologists were frankly pretty embarrassed by.

When Archimedes discovered the principle of buoyancy *("A floating object displaces its own weight in water")* he leapt out of his bath and ran naked through the streets of

Syracuse shrieking *Eureka*—which, as everyone knows, is Greek for "I found the soap." Guth was not in the bath when he had his Eureka moment, so he had to make do with writing SPECTACULAR REALIZATION on a sheet of paper and drawing a big box around the words.[9]

What he had seen was this: many of the problems with 'standard' Big Bang theory go away if you assume that the universe went through a brief period of extra-rapid expansion—"inflation"—in the first fraction of a second of time.

Of course, in one sense the universe is 'inflating' right now: it's still expanding. But what Guth meant was different. He meant that for a certain incredibly brief instant the dimension of space *expanded faster than the speed of light.*

Impossible?

No. Einstein's equations tell us that nothing can *traverse space* (get from A to B) faster than light. But relativity doesn't say anything about how fast A can get away from B—that is, how fast *space itself* can expand. So Guth's idea, however

----

[9] EUREKA MOMENTS AND GENIUS: We love the idea that Einstein was a lone super-mind surrounded by mouth-breathers. The truth is much more complicated: his work was profoundly original, but it might never have happened without the exceptional earlier insights of Bernhard Riemann, James Clerk Maxwell, Max Planck, and many others. So with inflation theory: it's a nice story that Alan Guth's single flash of insight joined up the cosmologically large with the sub-sub-atomically small. But this beautiful and fruitful pairing was already in the air. In fact it really became inevitable decades earlier, in the 1920s, when the great Indian astrophysicist Subrahmanyan Chandrasekhar (Chandra, for short) grasped that you need quantum mechanics to make sense of white dwarf stars. (See also the notes above on DNA's Secret History, and on Edwin Hubble.)

counterintuitive, doesn't violate any known physical laws.

There were also some good technical reasons to take inflation seriously, but it did sound a bit phony at first; it had the ring of a cleverly-engineered solution designed specifically to sweep problems under the carpet.

The shy phase didn't last. Like many good scientific theories, inflation led to specific predictions, huge grant proposals, and large, complicated, satellite-based experiments. The theory passed the tests brilliantly. That's when cosmologists started to mull over the possibility that some version of Guth's idea was probably *not* just clever mathematics but rather a description of our world.[10]

While the scientific community was still digesting this, and getting a bad case of heartburn as a result, along came Russian cosmologist Andre Linde.

Linde saw a problem with Guth's theory, and came up with a solution. The solution was inflation in a new form, which Linde calls "chaotic inflation." The new version was even more persuasive. And it had consequences even more ludicrous than the old one.

We have already seen that the *whole* universe may be bigger than the *observable* part. One consequence of any inflation theory is that it must be much, much bigger.

How much is 'much, much'?

Take a deep breath.

Inflation occurred only for an unimaginably small first fragment of the first second, (it *ended* at 10^{-35} of that first second), After that, expansion continued at slower-than-light speeds. This "inflationary epoch" (yes, they really call it that) was so brief that, even at light-speed, a Planck-length uni-

---

[10] For more on why inflation has become persuasive, see Appendix 4.

verse could only have gotten a teeny bit bigger. But inflation theory says it grew to the size of a baseball—many orders of magnitude bigger. According to various versions of Linde's theory, this means the radius of the universe *now* is 10^{100}, or possibly $10^{40,000}$, or possibly "ten to the ten to the twelve" centimeters.

Since all these numbers are perfectly insane, let's look first at the most insane of all. In the end, as you'll see, it doesn't make a whole lot of difference.

"Ten to the ten to the twelve" (10 raised to the power of 10^{12}) might not look very impressive to a non-mathematician. That's a pity, because in fact it's a number that makes a term like "mind-bendingly humongous" seem almost hysterically inadequate. And 'hysterical' is the word: once you start to understand "10 raised to 10^{12}", sobbing with horrified laughter is a perfectly reasonable response. "10 raised to 10^{12}" is, in the most literal sense of the word, *monstrous*. To borrow a phrase from Shakespeare, it beggars all description.

But hey, let's have a go.

For the sake of simplicity (and because, as you'll see, the units of measure don't matter a hoot either) let's forget all about whether we're talking centimeters, or light-years, or Tera-parsecs. Let's just say the "Linde Universe" is "10 raised to 10^{12}" times as big as our observable universe.

That's not 12 times bigger, or even "10 x 12 (= 120)" times bigger. Nor is it "10 raised to the 12" (10^{12} = one trillion) times bigger. A trillion times bigger—that would be headline stuff! But no. The Linde Universe is "10 raised to the '10 to the 12th power'" times bigger. That's *ten raised to the power of a trillion* ($10^{1,000,000,000,000}$) times bigger.

This is a number far, far too big to write down in ordinary notation. It's also far too big to make sense of, but maybe we can try to get a little sense of just how little sense of it we can get ...?

A quick refresher on the power of powers is in order here. Pick a seriously big number: say, the number of protons and neutrons in the observable universe. That's said to be somewhere around 10^{80}, which is 1 followed by 80 zeroes. Big. But remember that, the way powers of ten work, a *trillion* universes like that (10^{12} of them) would have only $10^{80} \times 10^{12} = 10^{80+12} = 10^{92}$ protons and neutrons.

So if you can imagine the difference in scale between (a) a single proton, and (b) a trillion of our universes—which, of course, you can't—you've not even taken the first step up the mountain towards imagining the scale of the Linde Universe.

Suppose instead that we attempt to get a grip on the idea by using a model. A really, really tiny model. Let's shrink our vast "observable universe"—all 100 billion galaxies or so—to the size of a grain of sand.

Or: let's reduce everything further, to the point at which the 10^{26} meter observable universe is the size of that speck-within-a-speck-within-a-speck, the proton.

Or—since extravagance seems to be in the air, and since an individual "string" is twenty orders of magnitude smaller again (remember all those people lined up between here and Sirius?)—let's reduce the observable universe to the size of one string.

This takes the observable universe from 10^{26} meters down to the Planck length, 10^{-35} meters, a precipitous fall through *all 62 orders of magnitude represented in this book.*

Where is this infinitesimally reduced universe situated?

Well, if we've reduced the surrounding Linde Universe by the same amount, to create a "Linde Miniverse" that's also 62 orders of magnitude smaller—

Instead of being $10^{1,000,000,000,000}$ times bigger than the observable universe, the Linde Miniverse is only

$10^{1,000,000,000,000}$ / 10^{62} times bigger.

Which is $10^{(1,000,000,000,000-62)}$ times bigger.

Which is $10^{999,999,999,938}$ times bigger.

Here's an admittedly paradoxical way to put this: At least from the human point of view, the Linde Universe is so utterly and absurdly gargantuan that shrinking it to a trillionth of a trillionth of a trillionth of a trillionth of a trillionth of its original size doesn't really make any difference to how big it is.

Of course, on some of Linde's models the whole universe is much smaller than this: a mere pathetic 10^{100} times bigger than the observable universe. In which case our string-sized micro-universe sits in a Linde Miniverse that's a mere pathetic $10^{100-62} = 10^{38}$ times bigger than the observable universe.

In ordinary language, that's *a hundred trillion trillion trillion trillion* times bigger.

Give up yet? Had enough?

It gets worse!

First: 10^{100} and $10^{1,000,000,000,000}$ are to some extent figures pulled out of thin air: the real power might be smaller or larger—or it might be, as Alexander Vilenkin and others have theorized, that space is simply infinite. (This isn't 'just a theory.' Some measurements of tiny irregularities in the Cosmic Microwave Background make sense if space is infinite and are hard to explain if it isn't.)

Second: Guth, Linde, and Vilenkin all doubt whether our universe, however big it may be, is all there is. On the contrary: inflation theory strongly suggests that our universe is just one "bubble" of spacetime among others—nothing more than a single fragment of fizz in the champagne glass of creation. If this is right, then there are other universes out there. Or out somewhere: these are not just other regions of space but in a deeper sense separate spaces, and there may be new ones fizzing into existence all the time.

In a new twist on the idea of the Creation, Guth has also suggested that, given a microscopic fraction of matter as a seed, and the right technology, an advanced civilization might be able to 'grow' new universes at will. As he says, with characteristic dark humor, our universe may have originated in someone's basement. It's rather disturbing to think that the God of the Creation might have been a minimally competent twelve-year-old in some other universe, and that *our* universe might be a runaway bench-top experiment.[11]

---

[11] THE LINDE UNIVERSE AND FOREST CONSERVATION: Try just printing out the "Linde number" in full. A quick calculation shows that with an eyestrain font your printer can squeeze about 10,000 zeroes onto a standard piece of paper. (You are using both sides, of course, in what turns out to be a doomed attempt to save trees). That's 5 million zeroes on a 500 sheet "ream." Which is 200 reams (a 10 meter stack—as tall as a house) per billion zeroes. But wait: that's just a thousandth of the work. Keep on churning: eventually you will have a stack of zero-covered paper that's a danger to intercontinental air traffic—it's higher than Everest.

Scale Omega

Are we there yet? Nobody knows.

Omega, the last letter of the Greek alphabet, is often used to stand for last things, or infinite things. The truth is, we don't have any better idea now of the ultimate scale of things than people did in the time of Galileo. Oh, we know more. But part of knowing more is knowing how ignorant we are. The philosopher Socrates said he was the wisest man in Athens, and people thought he was boasting. Then he explained that what he meant was just this: everyone he spoke to believed they knew things it turned out they didn't know; he was wiser only because he *knew that he didn't know.* (You can't win. The Athenians concluded that he was making fun of them, and gave him poison to drink.)

All in all, it seems unlikely that superstring theory and the Guth / Linde multiverse are the end of the road. Maybe the world around us is even bigger (and smaller) than living scientists have yet imagined.

That we have arrived at some kind of genuine 'ultimate' seems more probable for the superstring than for the Linde multiverse, since part of the very attraction of string theory is that (if it's right) it puts an absolute lower limit of 10^{-35} on meaningful scale—rather as Einstein put an absolute upper limit on meaningful velocity. The multiverse, on the other hand, is just the biggest thing anyone has ever thought of a theoretical reason for proposing.

But if the past is anything to go by, we have little reason to be confident that we 'finally know the truth' in either direction ...

Did I mention British physicist David Deutsch, who argues from a few uncontroversial facts about quantum mechanics to the conclusion that any *one* universe like ours must be, in fact, one of an interwoven set of about 10^{500} almost-identical parallel universes?

Or did I mention Swedish-American cosmologist Max Tegmark, whose calculations indicate that within "10 to the 10 to the 28" meters from us there must be a planet that's *identical to the Earth in every single detail of its entire history* except that your name is Obadiah?

Guess not.

OK, sorry, I'll shut up. It's a good place to end—and no better time to remember the epigraphs with which we opened our exploration:

The beginning of every science is the description and naming of phenomena.

Where the telescope ends, the microscope begins. Which (of the two) has the grander view?

If at first an idea is not absurd, then there is no hope for it.

I hope you enjoyed the tour. Welcome home!

Appendix 1: Some Random Notes I Couldn't Resist

Deep time

What do we mean by *"sauropods lived between 165 million and 65 million years ago"*?

Suppose it's lunchtime now. If the first sauropod walked the Earth at lunchtime yesterday, flowering plants came along to keep them company during the late afternoon, and the great Cretaceous-Tertiary extinction removed them from the scene in the early hours of this morning. About dawn (a bit too early to see grass evolve) the land-dwelling ancestors of whales headed back to the sea. Modern *Homo sapiens* wiped out its close cousin *Homo neanderthalis* 20 seconds ago, built the Pyramids at Giza three seconds ago, and landed on the Moon a few thousandths of a second ago.

Or let's put it another way. Suppose human civilization lasts as long as the sauropods—a hundred million years—and then someone writes a detailed history of the whole party. Such a book will run to 20 volumes of 500 pages each—and everything so far, from the invention of writing to global warming, will be on page one of Volume One.

Weird gunk and very weird gunk

Biologists long claimed that the "five kingdoms" of life divide into two "superkingdoms": the more primitive prokaryotes (organisms with a single cell that has no nucleus, i.e. bacteria) and the eukaryotes (everything else). But in the 1970s they discovered weird forms of "bacteria," the Archaea, that aren't really bacteria at all. Archaea tend to live in very strange places, like boiling hot springs, extremely salty water and deep undersea rock—places where nothing else, not even bacteria, can survive—and it has been said that genetically they are as different from bacteria as we are from spiders. The eukaryotes, including us and the spiders, probably evolved from them. Huge colonies of Lithotrophic (literally, "rock-eating") Archaea produced the methane (natural gas) that is one of our key energy sources. How huge? "SLiMEs" (Subsurface Lithotrophic Microbial Ecosystems) may make up as much as a third of all living things by weight.

Cellular slime molds—not to be confused with SLiMEs—were long believed to be fungi, but genetically many of them seem to be as unrelated to fungi as they are to plants or animals. Also called pseudoplasmodia, they are loose groups of individual cells that can communicate chemically and join into mobile masses that are rather like a slug. While feeding, they are essentially a single cell with thousands of separate nuclei—but they can reach a foot across. These damp chunks of goo can do spookily intelligent things, like crawl in a particular direction and even find the shortest route through a maze.

Catch my drift?

In 1912, German scientist Alfred Wegener came up with the theory that the continents were adrift—that they moved around and even split apart and rejoined over geological time. There was a lot of fossil and other evidence for this, but many geologists dismissed the idea. How could continents *drift*?! Wegener died in 1930, and it was not until a generation later that most scientists accepted his thesis. But don't stand at the shore waiting for an impact: continents drift at about the same speed as your fingernails grow.

Good reef

The individual animals that make up coral reefs are only a few millimeters across. But they link together (the linking part is called the coenosarc), and they produce *communities* of living organisms bigger than any other on Earth—and they provide a habitat so rich that they have been described as "the rainforests of the sea." The Great Barrier Reef, off Queensland in north east Australia, is almost as long as the Himalayas. Unfortunately the world's reefs are being killed off, rapidly and perhaps irreversibly, by ocean acidification.

A magnifying glass, only better

As early as the 1870s, it was clear that the wavelength of light put an absolute lower limit on the resolution of microscopes. Ever since the work of Louis de Broglie in 1924, which showed that electrons have wavelike characteristics, it has been obvious that you could 'shine' a beam of electrons through a specimen to create images of objects too small to be captured by the much larger 'waves' of visible light. This line of research has given us the original Transmission Electron Microscope (1931), the Scanning Electron Microscope (1965), and the Scanning Tunneling Electron Microscope (1981).

Optical microscopes have a resolution limit of about 0.2 micrometers. Electron microscopes have a limit around 0.2 nanometers—a thousand times better. Scanning tunneling electron microscopes can image, and even arrange, individual atoms.

Big waves, rogue waves, tsunamis

Waves of 10 to 12 meters (the height of a two-story house) are a nasty business when you're at sea. But there's now irrefutable evidence that centuries of mariners' stories about much larger "rogue" waves, which seem to rear up out of nowhere, are not the tall tales they were thought to be.

During a North Sea storm with 12 meter waves on January 1, 1995, the Draupner oil platform was hit by a single 26 meter (85 foot) wave; the "Draupner wave" has now become famous as the first rogue wave for which there was irrefutable proof. In the same year, while trying to avoid Hurricane Luis in the North Atlantic, the liner Queen Elizabeth II was struck by a 29 meter (95 feet) wave; the captain described "a great wall of water" that made it look "as if we were going into the White Cliffs of Dover." (In 1943, also in the North Atlantic, the original Queen Elizabeth was struck by a huge wave and sustained damage 28 meters above her waterline.) On April 14, 2005, the cruise ship Norwegian Dawn was hit by a "seven story" (20-22 meter) wave en route from the Bahamas to New York; a spokesperson for the cruise line said that "the sea had actually calmed down when the wave seemed to come out of thin air." During a week-long storm in the North Pacific in 1933, the Navy tanker Ramapo encountered one monster wave 34 meters (112 feet) high.

Such reports are rare, but that's not surprising. Encountering a 30 meter wave is roughly like having a ten-story office building collapse on your head. Even a 200 meter supertanker, meeting one of those at the wrong angle, could be rolled or torn apart and disappear without a trace. The *München* may have met this fate: one of the largest cargo ships ever, it sent out a garbled mayday message from the mid-Atlantic on December 12, 1978. No survivors and almost no wreckage were found.

In 2001, the European Space Agency's MaxWave project dedicated two satellites to a detailed observation of the ocean using synthetic aperture radar; they observed ten waves in the 25 meter range in just three weeks.

Automatic ocean sensors in the Gulf of Mexico, over which Hurricane Ivan happened to pass on the night of September 15, 2004, repeatedly measured waves up to 28 meters. A computer model suggested that, given the data from the sensors, there were probably waves in the area up to 40 meters. On the other hand, this should be treated with some skepticism: it was computer models that persuaded scientists that rogue waves didn't exist.

"Rogue" sea waves are very different from, but about the same size as, the worst tsunamis—monster waves caused by undersea earthquakes that come ashore with almost no warning. Many destructive tsunamis are 3-15 meters high—including many of the waves that caused such terrible destruction in the Indian Ocean on December 26, 2004. But on April 1, 1946, Scotch Cap lighthouse in Alaska's Aleutian Islands was wiped away like a paper toy in a 35 meter tsunami, the largest in U.S. history.

Rogue waves at sea, and coastal tsunamis, are mercifully rare. Even more unusual—but even bigger—are the local waves sometimes caused by landslides. On July 7, 1958, when the eastern wall of Alaska's Lituya Bay collapsed, it created a wave so big that, when it hit the opposite shore, it uprooted trees *five hundred* meters above the waterline.

A similar event may occur, on a vastly greater scale, when the volcano Cumbre Vieja collapses, as it probably will during its next major eruption. Cumbre Vieja is on La Palma in the Canary Islands. Some models predict that 500 cubic kilometers of rock will slide off its western flank into the Atlantic, generating thousand-meter waves locally. Those waves would dissipate to "only" 10-100 meters when, hours later, they devastate coastlines including West Africa, southern Britain, the Caribbean, and the entire eastern United States.

Little space invaders

Imagine you're looking up at the sky in the middle of the night. The Sun is 150 million kilometers beneath your feet, on the far side of the Earth, and its photons can't reach you. But the Sun is also pouring forth neutrinos. Neutrinos, predicted by Wolfgang Pauli in 1931 and given their name ("little neutral one") by Enrico Fermi—are strange, extremely light particles that pervade all of space and yet almost never interact with anything. The ones generated by the Sun pass right through the Earth as if it isn't there. Then they burrow up through the soles of your feet and launch themselves back into space through the top of your head, at the rate of 100 trillion per second. By the time you finish reading this sentence, the neutrinos that were passing through your brain as you began to read it are already out beyond the orbit of the Moon.

Long voyage(r)

This book tries to stick to the natural world, but it is impossible to resist noting the awe-inspiring achievement of NASA's Voyager missions. Voyager 1, the space probe launched on September 5, 1977, is now the most distant object humans have ever made. It has sped out of the solar system into interstellar space at 100,000 kilometers per hour, and is about 130 times as far away as the Sun—not eight light-minutes from us, but eighteen light-hours.

If you were a colossus in space, and your steps were as wide as the Earth, it would still take a million paces—several weeks of hiking—to reach where Voyager 1 is now.

Thanks to Carl Sagan, who had both the imagination to think of it and the sense to pester the project engineers who hadn't thought of it, in 1990 NASA instructed Voyager 1 to take one picture looking back over its shoulder: the first image of our solar system. Sagan's famous words are worth repeating: *"Look again at that dot. That's here. That's home. That's us. On it everyone you love, everyone you know, everyone you ever heard of, every human being who ever was, lived out their lives. The aggregate of our joy and suffering, thousands of confident religions, ideologies, and economic doctrines, every hunter and forager, every hero and coward, every creator and destroyer of civilization, every king and peasant, every young couple in love, every mother* and father, hopeful child, inventor and explorer, every teacher of morals, every corrupt politician, every "superstar," every "supreme leader," every saint and sinner in the history of our species lived there—on a mote of dust suspended in a sunbeam."

You can check on the progress of Voyager 1 and Voyager 2 at http://voyager.jpl.nasa.gov, the official mission website.

What makes this stuff *this* stuff?

Alchemists dreamed of turning lead into gold, but they were barking up the wrong cauldron; chanting spells and adding tincture of bat's wing just won't do the trick. You don't get gold by adding stuff to lead. What you need to do is remove three protons. Every atom of lead contains 82; gold contains 79.

The Periodic Table shows all the possible types of fundamental "stuff"—the 94 elements found in nature and the 20 or so that have been created in experiments. Each has its unique number of protons and electrons.

Neutrons are different: adding and subtracting neutrons just gives you a different "isotope" of the same element. It may have new properties, like being radioactive, but it's still fundamentally the same stuff. Deuterium, also known as 'heavy hydrogen,' is hydrogen with a neutron as well as a proton; tritium is hydrogen with two neutrons as well as the proton.

How I wonder where you are

How do we know how far away stars and galaxies are? The answer is: it's really, really difficult, but astronomers and cosmologists use many different methods, so that the numbers you get one way can be checked another way.

For nearby stars, the key method is parallax. In March, on one side of the Sun, the Earth is 300 million kilometers from where it is in September, on the other side. That makes nearby stars seem to shift relative to the background of very distant stars. (Stick out your thumb, then look through one eye at something behind it. Now hold your thumb steady and switch eyes. That's the idea.) If you measure the angles accurately, and know that the base of the triangle is exactly the Earth's orbital diameter (2 AU), it's easy to calculate the distance.

Enough accuracy is difficult; even for the nearest stars, the change in angle is similar to what you would observe if you compared the left and right edge of a small coin several kilometers away. And, the further out you go, the harder it gets—a fact illustrated by the much-studied star Betelgeuse. The European Space Agency's HIPPARCOS satellite measured the distance to Betelgeuse at 643 light years, *plus or minus 146 light years.*

For nearby galaxies, you can at least make a start with the 'brightest star' method. Assuming there's an upper limit on how bright a star can be, the very brightest stars in our galaxy are about as bright as the very brightest in nearby galaxies. Measure how much dimmer the extragalactic stars are and you can calculate how far away the galaxy is.

Supernovas and Cepheid variable stars are just two more 'standard candles' that can be used to estimate distances. How standard they really are is part of what we don't know.

Astronomers in the dark

Many authorities give a mass for the Milky Way that's around 10^{41} kilograms or 100 billion solar masses. But this is doubly misleading.

First, astronomers know that much of the universe's matter—about 85% of it—is "dark matter." Which is to say this: if you add up all the stars, all the dust, and all the stuff you can't see because it's behind something else, you find out that *most of the universe* must be something else. There's a lot of argument about what the dark matter is. Zillions of microscopic black holes? Giga-zillions of previously undetected subatomic particles? A truly exciting quantity of undiscovered chocolate? Your guess is as good as theirs. In the case of a galaxy like the Milky Way, "dark matter" calculations suggest that the galactic mass is nearer 2 trillion solar masses.

Look up "dark energy" and you will see that things are really a lot worse than that. When cosmologists did detailed calculations of the rate at which the expansion of the universe is slowing down, they discovered that it ... isn't. In fact it's speeding up—and this acceleration is driven by "dark energy."

Think of all the mass-energy in the universe as a crate of 100 apples. Just 5 are regular stuff, like people, parrots and protons. Just 27 more are the mysterious dark matter. The other 68 are the even more mysterious dark energy.

Far star
"Sanduleak 1987A" may not be much of a name, but this star is lucky to have a name at all. 'Nearby' stars are under a hundred light years away. This one isn't even in our galaxy—it's *150,000* light years away, in the Large Magellanic Cloud, one of our satellite galaxies. But it put on a big show: having been a 20-solar-mass blue supergiant, about 100,000 times as bright as the Sun, in February 1987 it became the first 'naked eye' supernova since the invention of the telescope. ("Kepler's Star" appeared in 1604.) In its death throes, Sanduleak 1987A briefly gave off the light of perhaps a billion stars—and, relative to theoretical predictions, astronomers considered this output something of a disappointment.

Hubble's hidden helpers
The fame, and the eponymous space telescope, are well

deserved: Edwin Hubble was probably the most important astronomer since Galileo. But his great work would have been impossible without the contribution of at least three other brilliant people who are virtually unknown. A decade before Hubble's great discoveries, his colleague Vesto Slipher began to find evidence that most "nebulae" (galaxies) were rushing away from each other. And soon after that the brilliant Russian physicist Alexander Friedmann—who seems to have understood Einstein's equations better than Einstein—claimed that if Einstein's theories were right the universe must be expanding.

Then there was another brilliant scientific mind working in even greater obscurity right alongside Hubble. Born with a terrible disability—two X chromosomes, instead of an X and a Y: yes, she was *female*—Henrietta Swan Leavitt was unable to get a job in science except as a lowly 'computer.' (In those days, the meaning of the word "computer" was "a person employed to do long tedious calculations.") A combination of hard work and superb scientific insight allowed her to unlock the secrets of the Cepheid variables, a class of stars that vary in brightness in proportion to their size. The importance of this discovery for our understanding of the universe was profound: it proved that previous ideas of the size of the universe were wildly, wildly wrong, and it showed how to map astronomical scales anew. If Leavitt had been a man, and had therefore done her work under the title she so richly deserved—"astronomer"—she would certainly have won a Nobel Prize. (See also the notes on DNA's Secret History, and on Eureka Moments.)

It's snail science

According to Einstein, traveling faster than light will never be possible, however fancy the rocket (see Appendix 3). But surely we can travel at *nearly* the speed of light?

Don't hold your breath. NASA's Juno spacecraft is currently the fastest object ever made. It's on its way to Jupiter at 30 km/s, or nearly 70,000 mph. But that's not 10% of the speed of light, or 1%, or even 0.1%—it's *0.01%, or a ten-thousandth*. A reasonable comparison: chasing a fast airplane by climbing onto the back of a snail.

Sure, one day we'll do better. But the most promising new technology on the drawing board, plasma propulsion, might perhaps take a spaceship to ten times Juno's speed. That's like exchanging the snail for a tortoise.

Are all quasars ghosts?

Light from a long distance away is light from a long time ago. We see the Sun as it was eight minutes ago, Sirius as it was 8.6 years ago, Betelgeuse as it was in the Early Middle Ages, and the more distant quasars as they were before the solar system formed. This line of thought carries us to a strange conclusion: there may be no quasars. Even the closest that we observe are very distant—nearly a billion light years away—and they are more common the further out (the further back in time) we look; in fact most are more than ten billion light years away. From this, we can infer that they were common in the early universe, but that a time came when no more were formed and the existing ones started to die out. If so, it may be that each quasar our telescopes "see" is just a ghostly message from the past—a note in a bottle from a sailor who drowned long ago.

Ah! That clarifies everything!

One of the weirder aspects of string theory is that it posits several extra, minutely "wrapped," spatial dimensions. We only experience three spatial dimensions, because we're too big to notice the seven or so sub-sub-sub-atomic ones. Strings, on the other hand, are just the right size, and they inhabit a family of multidimensional mathematical objects called Calabi-Yau spaces. A Calabi-Yau space is usually represented as something like a ball of string, but the squiggles are just an attempt to represent on two-dimensional paper a three-dimensional aspect of a ten-dimensional object. Or, as string theorist Bryan Greene helpfully says of such an illustration in one of his books: *"For the mathematically inclined reader we note that this particular Calabi-Yau space is a real three dimensional slice through the quintic hypersurface in complex projective four-space."*

Small change

If the number of protons in the universe is about 10^{80}, the number of atoms in the universe is also about 10^{80}. Three quarters of all atoms are hydrogen and each hydrogen atom has one proton. Most of the rest of the universe is helium, but the two protons in each helium atom are not nearly enough to get us all the long, long way from 10^{80} to 10^{81}.

Just a flash

The speed of light is now the fundamental standard from which all length measurements are derived. The meter was originally defined—by a committee of France's Academy of Sciences in 1790—as one ten millionth the length of an arc running from the North Pole to the Equator. In 1983 it was redefined as the distance light travels in a vacuum in 1/299,792,458th of a second (about 3.3 billionths of a second).

The briefest human-made events are bursts of light a bit less than a femtosecond (10^{-15} seconds) long, the duration of the shortest currently feasible laser and x-ray bursts. A femtosecond is a very, very, very small amount of time. There are a million billion femtoseconds in a second. If you slowed time down so that a femtosecond lasted one second, one second would last 35 million years.

The briefest known natural events, such as the time it takes an electron to move from one "shell" or "orbit" of an atomic nucleus to another—are around one yoctosecond. That's 10^{-24} second, or a billionth of a femtosecond. A yoctosecond bears about the same relation to a microsecond as a microsecond does to the age of the universe.

Eternal life, sort of

Relativity theory predicts *time dilation*—the faster you go, the slower your time seems to pass from the viewpoint of the twin you left on Earth.

This doesn't mean that interstellar travel will be boring, though it certainly will be. Time will seem entirely normal on your long dull journey. But the flow-rate of your time, from the viewpoint of an observer, tends to zero as the difference between your speeds approaches *c*. If you travel at 99% of *c*, your earthbound twin will seem to age 7 hours during your lunch hour (and, paradoxically, you will seem to age 7 hours during hers.) But at 99.99% of *c*, the discrepancy is 70 hours per hour; at 99.99999999% of *c*, we're talking years.[12]

Experiments have confirmed this precisely, and it implies that photons in the vacuum of space, traveling at precisely the speed of light, are in a sense immortal. Imagine you could become a photon and do a lap of the Milky Way at speed *c*. When you returned from your Very Grand Tour (in 300,000 A.D.) your friends and family would be long gone; in fact, the Earth and its biota might be unrecognizable. But you would not have aged by a single millisecond.

This means that time travel *into the future* is no science

---

[12] PARADOXICALLY ... Here's the notorious Twins Paradox, in a nutshell: "When you travel to the stars, and I stay home, my clock runs slow relative to yours, *and vice versa*. So why am I older than you when you return?" The solution is ... too complicated to explain here. But be sure to read Robert Heinlein's classic 1956 riff on the subject, *Time for the Stars*.

fiction myth: all you need is a ship that will take you to a large percentage of c. Unfortunately, the only way to return to your own time is to design an engine that burns science fiction.

Appendix 2: Numbers and Names

10^{26} m   is 10 billion light years, halfway across the observable universe

10^{25} m   is one billion light years, the distance to nearby quasars

10^{24} m   is 100 million light years, the size of our galactic supercluster

10^{23} m   is 10 million light years, the thickness of the Great Wall

10^{22} m   is one million light years, half the distance to the Andromeda galaxy

10^{21} m   is 100,000 light years, the length of the Milky Way

10^{20} m   is 10,000 light years, the length of the Large Magellanic Cloud

10^{19} m   is 1,000 light years, the distance to the Horsehead Nebula

10^{18} m   is 100 light years, the diameter of a large globular star cluster such as Omega Centauri

10^{17} m   is 10 light years, the width of a typical globular star cluster or the Crab Nebula, and a bit more than the distance to Sirius

10^{16} m   is one light year, an envelope big enough for the "far solar system" including the Oort Cloud

10^{15} m   is a trillion kilometers

10^{14} m   is 700 AU, a bit less than the furthest point from the Sun (aphelion) of the highly eccentric planetoid Sedna

10^{13} m   is 70 AU, an envelope big enough for our "main" solar system including the orbit of Neptune

10^{12} m   is 7 AU, a billion kilometers, one light hour, twice the orbit of Mars—or the diameter of a supergiant star like Antares or Betelgeuse

10^{11} m   is 0.7 AU, the orbit of Venus or the diameter of the blue supergiant Rigel (the orbit of Earth is 1 AU by definition)

10^{10} m   is 10 million kilometers, the diameter of a small red giant star

10^{9} m   is one million kilometers, the diameter of stars like the Sun (the Sun's diameter is 1.4×10^{9} m)

10^{8} m   is 100,000 kilometers, a bit less than the diameter of Saturn

10^7 m   is 10,000 kilometers, the diameter of Venus (Earth's diameter is 12,700 kilometers)

10^6 m   is 1,000 kilometers, the length of Italy or California

10^5 m   is 100 kilometers, a bit bigger than Tycho, the most obvious crater on the Moon

10^4 m   is 10 kilometers, the height of Everest and the size of a small neutron star

10^3 m   is one kilometer, the height of Angel Falls

10^2 m   is 100 meters, the height of a large iceberg

10^1 m   is 10 meters, the height of a Brachiosaurus

10^0 m   is one meter, the height of a human child (or an adult sitting down)

10^{-1} m   is 10 centimeters, the size of a small rodent or a very large insect

10^{-2} m   is 1 cm, the size of many familiar insects and the smallest reptiles and amphibians

10^{-3} m   is 1 mm, the size of a small ant or a large mite

10^{-4} m   is 1/10 mm, the size of a human egg and the resolution limit of the human eye

10^{-5} m   is 1/100 mm, the size of a cell

10^{-6} m   is 1/1000 mm, a millionth of a meter, or one micron (μ), the size of a bacterium

10^{-7} m   is 1/10 micron, the size of a typical virus (and the resolution limit of a light microscope)

10^{-8} m   is 1/100 micron, the width of a bacterial cell wall or the size of a very small virus

10^{-9} m   is 1/1000 micron, a billionth of a meter, or one nanometer (nm), the size of a Buckeyball and the resolution limit of electron and scanning probe microscopes

10^{-10} m   is 1/10 nanometer, the size of an individual hydrogen atom

10^{-11} m   is 1/100 nanometer

10^{-12} m   is a trillionth of a meter, or one picometer

10^{-13} m   is 1/10 picometer

10^{-14} m   is 1/100 picometer, the size of a uranium nucleus

10^{-15} m   is one femtometer, the size of a proton or neutron

10^{-16} m   is 1/10 femtometer

10^{-17} m   is 1/100 femtometer

10^{-18} m   is one attometer, possibly the size of an 'up' quark?

...

10^{-22} m   is a tenth of a zeptometer, possibly the size of a 'top' quark?

...

10^{-35} m   is the Planck length, the fundamental unit of spatial extension and possibly the size of an individual superstring

Appendix 3: The Tale of Zippy the Tourist (Einstein Sheds Light on Time)

Contrary to popular belief, Einstein's theories aren't about the speed of light. They're about the speed of *electromagnetic radiation*. That includes the radio waves that bring music to you, the microwaves that heat your lunch, and the narrow band of waves ("visible light") that you can detect directly because of the weird pair of jelly-like detectors that evolution has installed in the front of your head.

All this radiation is fundamentally the same: photons. And all these photons travel (at least in the vacuum of space) at a whisker less than 300 million m/s. (That's New York to Los Angeles and back in a thirtieth of a second.)

Einstein grasped that this isn't just any old high speed. In the framework of relativity theory, the c in $E = mc^2$ is a fundamental physical constant and the fastest speed that makes sense in our world—the rate of propagation of pure energy. ("c" is for *celeritas*, the Latin for speed.)

But ... the fastest speed that *makes sense*? Why not just rev the engine a little bit harder?

Good question.

According to Einstein the problem is this. You can accelerate *things* (molecules, pies, astronauts, galaxies) until their speed gets closer and closer to that of light—all you have to do is use a lot of energy. But all these *things* are made of matter (they have mass), and Einstein showed that for them, weird as it sounds, increasing speed increases what you might call "effective" mass, so that the faster you're already going, the more energy it takes to go a bit faster. (An object has 'rest mass' and 'relativistic mass'—it's the relativistic mass that increases with speed.) The increasing mass due to increasing speed is too tiny for an observer to detect when you run for the bus, throw that pie, or take a rocket to the moon. But lob that pie at 30 miles per hour and its relativistic mass will increase by some fraction of a trillionth of a gram. Adding energy to a thing (by making it go faster) makes it harder to make it go faster still.

Now, when a *thing* gets to about half the speed of light, that extra mass becomes significant. In fact, at that speed the additional relativistic mass roughly equals the rest mass. By 90% the speed of light, the relativistic mass dwarfs the original mass. And by 99.99%, the graph of total mass (and therefore of the amount of energy you have to put in to go any faster) is climbing steeply towards infinity. (The relativistic mass of protons, pushed to very near c in the Large Hadron Collider at CERN, reaches several thousand times rest mass.) A mouse traveling at a bit less than c will have the relativisitc mass of a whale. Pour in more energy and it will go faster—a bit—but at the new, higher speed it'll have the relativistic mass of a planet ... a star ... a galaxy. Eeek.

Still, at least you can accelerate *things* in this way by pumping in more energy. What you can't do is pump energy into energy. Energy is as energetic as it can be, and pure energy (without any mass at all) is what photons are. They don't need accelerating to c with extra energy, because they *are* energy, and they always, by their very nature, travel at c, at least in a vacuum.

OK, this is a simplification. It's true that (say) blue photons are more 'energetic' than red photons. That's why a young, hot star like Vega is blue, while an old, cool star like Betelgeuse is red. But photons carry extra energy by having a shorter wavelength—never, not even a teeny bit, by moving faster.

Yes, but why?!?

Well, perhaps it was misleading to begin by saying that

'c' is the speed of electromagnetic radiation. Einstein didn't measure the speed of radiation first, and then announce that nothing would ever be discovered going faster. What he saw was that, because of the the relationship between the ideas of *cause*, *effect*, and *simultaneity*, a "no maximum" universe is nuttier than the inside of a squirrel.

Imagine this. It's January 1, 2025. You just received a radio message that was transmitted ten years ago by Zippy, your pen-creature, who lives 10 light years away on a planet orbiting the star Epsilon Eridani. Her message goes like this:

"January 1, 2015. Greetings, Earthling. I'm building a spaceship that will be capable of doing ten times the speed of light. It also has lowered suspension, twin mufflers, flame decals, and a little plastic Jesus on the dash. I estimate that I'll finish tinkering, and actually start my journey, in ten years—just about the time you receive this message! Heat up some soup and leave the light on."

You double-check the dates and notice that, sure enough, Zippy's radio message has been obedient to the known laws of physics: exactly ten years (time) to travel the ten light years (distance) from Epsilon Eridani. So you quickly swing your telescope into the constellation Eridanus, to the right of Orion, hoping to watch your friend's fancy new machine make its journey.

But you can't watch it.

You absolutely, absolutely, absolutely can't watch it.

Not now, not ever.

If Zippy blasts off on schedule, on January 1, 2025, she will make the trip to Earth in one year, arriving on January 1, 2026. But any light bouncing off her craft in the direction of your telescope—and also the photons generated by her daily radio diary, Starbook updates, Galctifeeds, and absolutely any other attempt at communication whatever—will fall far behind in her wake, traveling at c. (Think of the analogy that you can't hear a plane coming if it's going faster than the speed of sound.)

Not only will none of her communications get to you before she does. The stuff she broadcasts late in her journey has less far to go, so it will get to you *before* what she broadcasts early in her journey.

What does this mean? It means something remarkably creepy. You will never observe her journey from Epsilon Eridani to Earth. Instead, Zippy will magically materialize out of nowhere on January 1st, 2026—*Greetings Earthling! Take me to your bathroom!*—and then you (and she) will get to spend the next ten years watching in horror as she recedes, talking backwards and growing younger by the minute, back to where she came from.

So: as long as she's moving faster than light, she's *not* moving faster than light: she's moving *slower than light and backwards in time.*

Physicists (who apparently have too much of this mysterious "time" stuff on their hands) have even put this into a limerick:

There was a young lady named Bright
Whose speed was faster than light.
She went out one day
In a relative way
And returned on the previous night.

Stephen Hawking has said we can be confident that there is no time travel. His argument is simple and quite powerful: where are the tourists? Using a similar argument, we might say: there is no faster than light travel, since we don't observe anything moving backwards in time. We can safely assume that Zippy's rocket science is less advanced than she claims.

Appendix 4: Gravity is Waving— and Everything is Nothing

By wild coincidence, I finished this book just as news came in confirming the existence of Einstein's predicted gravity waves. That in turn is a strong reason for betting (a) that inflation theory is true, and (b) that Alan Guth and Andrei Linde have a Nobel Prize in their future.

Take a look at the main cover image of this book—a map of the Cosmic Microwave Background created by the Wilkinson Microwave Anisotropy Probe. That's the light emitted 380,000 years after the Big Bang, at the time of "photon decoupling," when light first streamed unimpeded through space. The WMAP images were done to show the "anisotropy" (variation) in the heat energy of the CMB. But astronomers using the BICEP2 telescope at the South Pole have detected a telltale "swirling" or polarization hidden in those patterns. (BICEP stands for Background Imaging of Cosmic Extragalactic Polarization.) These swirls are probably caused by primordial gravity waves—and they would be nothing like big enough to detect without having been magnified first through cosmic inflation.

BICEP has not just detected gravity waves and therefore confirmed inflation, though. Because a crucial ratio r in the data is significantly bigger than expected, some inflation models are strongly favored and others seem to be ruled out.

Without going into the technical details about that, it's worth pausing to be suitably amazed by this: a careful look at some very old light really does give us clear experimental evidence in favor of some theories (and against other theories) about what happened to the universe 10^{-36} seconds after the beginning of time.

Now look at the cover image again. It's a picture of the largest structures in the current universe—but also of quantum fluctuations. Inflation "imprinted" quantum effects onto the larger universe it created, so galaxies and galaxy clusters are in effect magnified Planck-scale effects. *The very largest and smallest scales are the same thing.*

And that takes us close to another strange idea that flows out of inflation theory. All the mass-energy of ordinary matter, plus dark matter, plus dark energy, is positive. But all gravitational energy is stored ("potential" or negative) energy —and the two exactly cancel out.

So the universe doesn't have to "come from" anywhere.

The net mass-energy of the universe is zero.

Everything is *nothing*, showing off.

To The Reader: Mistakes Were Made

Dear reader,

Mistakes were made, somewhere in here, no question. Plus, things just get out of date—some facts I quote have changed repeatedly while I was working on the book.

Please don't gripe. Instead, send me a note at richard@richardfarr.net. Put *You Are Here* in the subject line, and as much detail and documentation as possible in the body. I'll add corrections as soon as I can.

Thank you thank you thank you!

Check out my website, buy my other books, and have a nice universe.

Richard

richard@richardfarr.net

www.richardfarr.net

More to Explore

Several films explore the idea of scale. One of the most famous is *Powers of Ten,* created in 1977 by Charles Eames. It's worth watching anyway, but especially to note where it stops and what it leaves out: we've learned a lot since then. Also look up *Cosmic Zoom, Cosmic Voyage, The Known Universe,* and the videos you get when you search (for example) 'star size comparison.'

The web is full of good "scale" articles, diagrams and videos—see for example *How Big Are Things?* at vendian.org. And don't miss the very cool interactive visualization by Cary Huang at scaleofuniverse.com.

About the Author

Richard Farr's first book, *Emperors of the Ice: A True Story of Disaster and Survival in the Antarctic, 1910-13*, tells of the legendary 1911 'Winter Journey' across Ross Island, a near-suicidal bid to collect Emperor penguin eggs and test Ernst Haeckel's theories about natural selection. Available both as a heavily-illustrated hardback and as an e-book, *Emperors* has won several awards, including a starred 'Outstanding Book' designation from the National Science Teachers' Association. Reviewers have described it as 'spellbinding,' 'enthralling,' and 'so gripping you will not want to put it down'.

Richard's novel *The Truth About Constance Weaver* (Fall 2014) is a mystery—or an anti-mystery—about obsession, art forgery, madness, and murder.

The Fire Seekers (December 2014) follows an ordinary American teen and the half-Scottish, half-Chinese genius who is (and isn't) his twin sister, as they try to puzzle out the real origin of civilization while staying one step ahead of the mysterious and terrifying Architects.

Projects in the works include a sequel to *The Fire Seekers*, a memoir (*What I Expected*), and a middle-grade novel set in the thirteenth-most-boring place in the world (*A Plague of Frogs*).

www.richardfarr.net